THE MODERN PRIMARY SCHOOL HEADTEACHER

—— Graeme Kent ——

Books for Teachers
Series Editor: Tom Marjoram

KOGAN PAGE

To the pupils, staff, parents and governors of St Thomas' Primary School.
This is how it might have been – if I had got it right the first time!

© Graeme Kent, 1989

All rights reserved. No reproduction, copy or transmission
of this publication may be made without written permission.

No paragraph of this publication may be reproduced, copied or
transmitted save with written permission or in accordance with
the provisions of the Copyright Act 1956 (as amended), or under
the terms of any licence permitting limited copying issued by
the Copyright Licensing Agency, 7 Ridgmount Street, London
WC1E 7AE.

Any person who does any unauthorised act in relation to this
publication may be liable to criminal prosecution and civil
claims for damages.

First published in 1989 by
Kogan Page Ltd
120 Pentonville Road, London N1 9JN

British Library Cataloguing in Publication Data

Kent, Graeme, *1933–*
   The modern primary school headteacher – (Books for
   teachers)
   1. England. Primary schools. Head teachers. Role
   I. Title   II. Series
   37212' 012' 0942

   ISBN 1-85091-864-3

Typeset by DP Photosetting, Aylesbury, Bucks
Printed and bound in Great Britain by
Biddles Ltd, Guildford

Contents

Introduction	**9**

1
The Headteacher's Job — 11
The style of the headteacher — 12
The duties of the headteacher — 14
The responsibilities of the headteacher — 15
Summary — 18
Self-assessment — 19

2
School Organisation — 21
Organisation of time — 21
Organisation of the timetable — 24
Organisation of the headteacher's time — 27
The school day — 29
Attendance of children — 29
Admission numbers — 30
Data Protection Act, 1984 — 30
Summary — 31
Self-assessment — 31

3
The Children — 32
The headteacher's image — 32
The headteacher as a teacher — 36
Discipline — 40
Record keeping — 51
Organisation — 52
Preparing children for secondary education — 55

Child abuse   56
Police and the school   56
Supervision after school   57
Summary   58
Self-assessment   58

4
The Staff   61

Recruitment of teachers   61
Deployment of teachers   67
Treatment of teachers   69
Appraisal of teachers   73
Training of teachers   76
Non-teaching staff   82
Dismissal of teachers   84
Staff facts   85
Summary   87
Self-assessment   87

5
The Parents   89

Drawing parents into the school   89
Communicating with parents   91
Involving parents in the education of their children   96
Parents as policy makers   101
Open evenings   103
Parent-teacher associations   104
Dealing with a bereavement   105
Parental facts   106
Summary   108
Self-assessment   108

6
The Governors   109

Appointment   110
Helping governors to carry out their duties   111
Governors' meetings   117
Special problems   121
Governors' facts   121
Summary   122
Self-assessment   123

7
Finance · 124
Local management in schools · 124
Fund raising · 126
The school fund · 129
Financial facts · 130
Summary · 130
Self-assessment · 130

8
Building and Plant · 132
The school building and the ethos of the school · 132
The building · 133
Plant · 135
Summary · 135
Self-assessment · 135

9
The Curriculum · 137
Designing the curriculum · 137
Children with special needs · 139
Sex education · 139
Religious education · 139
Assessment and teaching · 140
Summary · 140
Self-assessment · 140

10
Public Relations · 142
Publicising the school among the local community · 142
Publicising the school over a wider area · 150
Public relations with the local education authority · 152
Summary · 156
Self-assessment · 156

Further Reading · 158
Index · 161

Introduction

No one has ever claimed that being a headteacher is easy. We are expected to be many things – managers, educationalists, conciliators, facilitators, trainers, publicists, technologists and building inspectors, to name just a few.

Getting the mix right has defeated some of the world's noblest intellects. One 19th-century pedagogue complained that only Moses had suffered more than he had, and when the famed Swiss educator Pestalozzi finally abandoned his own efforts to run a school for deprived children, he wrote despairingly that for years he had been 'living like a beggar ... in order to teach beggars to live like men'!

The roll-call of the great and the good who have been unable to come to terms with the problems of headship is a long one. Froebel the devisor of the kindergarten, Robert Owen the social reformer, and Benjamin Franklin the statesman and inventor, were only a few of the theorists who opened schools only to see them crumble about their ears. If they had trouble organising a school, and this before the additional burdens of local management of schools, staff appraisal and parental involvement, what hope is there for the rest of us?

Happily I can suggest there is quite a lot. It is true that the headteacher's job has changed almost beyond recognition over the last few years and is continuing to expand. Today we have more duties, more responsibilities, more problems and probably more criticism than we have ever had. There is a wealth of new legislation to be understood, fresh methods to be mastered, and more accountability than most of us had ever dreamed possible.

Yet there is a brighter side to all this. For the first time since I entered primary-school teaching more than thirty years ago, the headteacher's tasks are being clearly defined. There should no longer be any excuse for not knowing what we are doing. The challenge is to do it properly.

This book is intended to act as a guide for all primary school headteachers, both new and experienced. It sets out the tradecraft of the headteacher's job in what is becoming an entirely new education service for us all. It is based on experience as a head and teacher in over a dozen primary schools, and discussions with literally hundreds of teachers and headteachers.

If you follow the advice offered it should help you to organise and run your school smoothly and efficiently. This will leave you time to make that personal contribution to the school which is the main function of all good headteachers. It should provide you with a number of short-cuts to making your school a good one in the terms by which, rightly, these matters are judged in the community's opinion.

Throughout the book, heads and teachers are referred to as 'she', while children are called 'he'. This is to avoid the clumsy use of he/she.

The book will not, I am afraid, automatically make you the district's outstanding headteacher. That will still be up to you personally. The craft is there for the learning. The art we must supply from within ourselves.

Chapter 1
The Headteacher's Job

The headteacher's job is to manage without, wherever possible, giving the impression of doing so. This has to be accomplished despite the fact that the elements which have to be controlled in the average primary school are varied and volatile.

Two studies made in 1988 emphasise the increasing stress of being a modern primary-school headteacher. A survey conducted by the National Association of Headteachers found that three out of four heads took early retirement, and that almost half of these did so because of the strain of the work. A Manchester Polytechnic investigation concluded that headteachers suffer greater stress than most managers, due to overwork, difficult relationships with staff, and coping with resources. Such pressure is something we all have to accept. Ours is a job without obvious parameters, but we owe it to ourselves and to our schools to ensure that our work is regulated and organised.

The sheer scope of a headship can be one of its pleasures. There are not many professions which carry with them a positive obligation to be academic, entrepreneur, accountant, athlete, editor, cheer-leader, and painter and decorator, all in the course of a single term. When added to these is the fact that in primary schools we are for the most part working with boys and girls who are still friendly, charming and responsive, the job has its own particular bizarre charm.

A considerable part of a headteacher's task consists of reacting to events. To do this successfully is an art in itself, but our main function is to be in charge of our schools. In today's climate this can be quite a subtle and sophisticated process. The managerial style best suited to the modern primary school is that of an informed headteacher who unobtrusively but firmly, brings her influence to bear on all aspects of school life. This means that as headteachers we must influence children, staff, parents, governors, the community, finance, the curriculum, administration and even time. To bring our schools to maximum effectiveness, we have much to do and many people to take with us.

The style of the headteacher

Schools have always taken their tone from their headteachers. The importance of the headteacher was recognised by Her Majesty's Inspectors in *Ten Good Schools* (HMSO, 1977) which stated that the one thing these establishments had in common was outstanding leadership. What makes a good headteacher is less easy to pinpoint. Generally it is a matter of style, of energy allied to foresight, personality, compassion and tough-mindedness. *Ten Good Schools* found that the successful heads had 'qualities of imagination and vision, tempered by realism'. They had the capacity to communicate specific educational aims to staff, pupils and parents 'to win their assent and to put their own policies into practice'. They were accessible, good-humoured and had a sense of proportion.

These qualities are revealed in different ways among headteachers. Some are extroverts, others Machiavellian, a number are 'nuts-and-bolts' mechanics working behind the scenes. We all have different qualities to bring to our schools. It would be a grey and dismal education service if all headteachers were to emerge from the same mould.

We should, however, examine the effect that our styles of leadership are likely to have on others. Do our colleagues, governors and the parents perceive us in such a way that they are likely to respect our opinions? We should ask ourselves how other people see us in our self-appointed roles.

MAIN TYPES OF HEADSHIP
The Teacher Says firmly, 'I am the head *teacher*'. Enjoys contact with children, timetables herself extensively. Often has to be dug out of a classroom to deal with problems.
Advantages: knows the children well, usually popular with colleagues.
Drawbacks: tends to lose overall control of the school, little time to spare for parents and other visitors, paperwork piles up.

The Educationalist Believes that her main task is to control the curriculum. Attends many courses, plans timetables meticulously, calls frequent staff meetings to implement syllabuses.
Advantages: knows where the school is going in educational terms.

Drawbacks: can lose touch with teachers and children, sometimes favours theoretical over practical aspects of running school.

The Manager Runs a tight ship. Never happier than when planning organisation of school, drawing up flow-charts, etc.
Advantages: school runs smoothly, teachers well backed-up, plenty of resources.
Drawbacks: sometimes loses sight of educational objectives of school.

The Salesperson Believes in putting the school on the map. Launches eye-catching schemes to publicise school. Always good for a quote in the local newspaper.
Advantages: gives the school a definite identity, often helps prevent falling rolls.
Drawbacks: sometimes regarded with suspicion by colleagues, if overdone can turn school into a bit of a joke.

We would all hope to include aspects of these main types of headship into our leadership style because each one had definite merits, although it is likely that we will incline towards one of them. There is nothing wrong with this, as long as we make a conscious effort to include strands of the other three into our work.

EFFECTS OF OUR LEADERSHIP ON THE STAFF
It is important to remember that as heads we are leading a number of disparate individuals, each with particular likes and dislikes. To run a school successfully we have to gain the active support of up to twenty men and women. It is our task to weld these individuals into a team, remembering that a style of headship that may be accepted by one teacher will not go down well with another.

This was brought home to me in my first year of teaching. The headmaster of my inner-city school was a remote man, seldom seen about the corridors. As a young probationer I saw nothing wrong with such out-of-touch and ineffectual leadership. Indeed, I rather welcomed it while I made my quota of mistakes in the seclusion of my classroom.

It was not until one morning assembly that I became aware that not all my colleagues accepted such hands-off leadership as equably as I did. This was brought to my attention when the middle-aged deputy headmaster standing next to me was unable to control himself any longer. Nudging me in the ribs he indicated the headteacher pontificating on the platform. 'Look at him,' hissed the deputy venomously. 'Just a big white cowboy on a big white horse.'

In this case, a style of leadership which was acceptable to an inexperienced teacher was regarded as unnecessarily pompous and annoying by older members of the staff.

The attitude adopted by the headteacher towards the staff forms an important part of the ambience of any school. Most research indicates that there are four main types:

'I'm in charge.' The headteacher is autocratic, hands down decisions, checks up on staff but does not allow them any initiative.

'I love you, but I know best.' The headteacher is friendly and considerate but takes all major decisions.

'Well done, everybody.' The headteacher develops a happy family atmosphere in the school, gives the teachers a great deal of autonomy, initiates little and checks up hardly at all.

'Let's get the show on the road.' The headteacher consults the staff fully, expects teachers to take a full part in decision making and the setting of standards.

Again, most headteachers will come into more than one of these categories, but with an inclination towards one of them.

The duties of the headteacher

No matter which method or combination of methods the headteacher adopts, there are a number of duties which must be performed.

- Organising, managing and controlling the school.
- Consulting the local education authority (LEA), teachers, governors and parents.
- Formulating the overall aims and objectives of the school and policies for their implementation.
- Helping to select staff.
- Deploying and managing staff.
- Maintaining relationships with unions representing staff.
- Determining, organising and implementing a suitable curriculum.
- Keeping under review the work and organisation of the school.

- Evaluating standards of teaching and learning in the school, and maintaining proper standards of professional behaviour.
- Supervising and participating in any agreed national framework of staff appraisal.
- Providing information about the work and performance of staff where this is relevant to their future employment.
- Ensuring that the progress of pupils is monitored and recorded.
- Devising and implementing a policy for the pastoral care of pupils.
- Being responsible, in accordance with the governors, for good behaviour and discipline among pupils.
- Arranging for parents to receive regular information about the school curriculum, the progress of their children, and other matters affecting the school, to promote common understanding of its aims.
- Promoting effective relationships with persons and bodies outside the school.
- Advising and assisting the governing body of the school.
- Providing for liaison and co-operation with officers of the maintaining authority.
- Maintaining liaison with other schools and further education establishments with which the school has a connection.
- Allocating, controlling and accounting for financial and material resources pertaining to the school.
- Organising security and supervision of school buildings.
- Participating in any national agreements for appraisal of headteachers.
- Participating in the identification of areas in which the head would benefit from further training, and undergoing this training.
- Arranging for the deputy headteacher or other suitable person to take the head's place when she is absent from the school.
- Participating, to such an extent as may be appropriate having regard to her other duties, in the teaching of pupils, including the provision of cover for absent teachers.

The responsibilities of the headteacher

The more planning and consultation that a headteacher can do, the better it will be for the school. Although in reality a great deal of our work will be done on the run, it should remain one of our main responsibilities never to lose sight of the long-term objectives of the school, however great the pressure on us. We should always be able to look inwards at the school and its needs, and outwards to its place in the community. There is a very real obligation to share our hopes and aspirations with our colleagues, but it is not realistic to expect them to be as enthusiastic about long-term projections. Even the most idealistic teacher will usually be too busy with her class and other immediate duties to wonder about the long term. That is one of the prerogatives of the headteacher.

Leadership

We have a responsibility to maintain leadership of the school, no matter how many times we may be distracted by other problems in the course of a day. It is a part of our image to appear to be in charge of any situation, no matter how many alarm bells may be ringing in our skulls. Our apparent imperturbability should not mean that we are ignoring the situation. If the headteacher can appear calm and in control, yet always concerned and supportive, despite constraints of finance, time and limited resources, then the rest of the school will take its tone from such sensitive use of authority.

The importance of the head's role as a leader has been emphasised again and again in studies of primary schools. Between 1980 and 1984, the Inner London Education Authority carried out its celebrated *Junior School Project* (ILEA Research and Statistics Branch, 1986). In the course of this study, fifty schools were studied and almost two thousand young children examined. The report concluded that there were twelve factors within the control of the headteacher and her colleagues which could be identified in all effective junior schools:

1. Purposeful leadership of the staff by the head
2. Involvement of the deputy head
3. Involvement of the teachers
4. Consistency of teaching among teachers on staff
5. A structured and organised day
6. Intellectually challenging teaching
7. A work-centred environment
8. A limited focus of subjects within a teaching session
9. Maximum communication between teachers and pupils
10. Thorough record-keeping
11. Parental involvement
12. A positive ethos within the school

Aims of the School

It is the duty of the headteacher, in consultation with the governors, to draw up and publish the specific aims of the school. These could be included in the school brochure or the headteacher's report to the governors, or annually in the governors' report to the parents. Some schools have them typed and framed and displayed on one of the walls in the entrance hall or the head's office.

These aims will vary from school to school, but a Department of Education and Science (DES) Green Paper on education published in 1978, provided a list of eight objectives which are used in many primary schools:

1. To help children develop lively, enquiring minds, giving them the ability to question and argue rationally, and to apply themselves to tasks.
2. To instil respect for moral values, for other people and for oneself, and tolerance of other races, religions and ways of life.
3. To help children to use language effectively and imaginatively in reading, writing and speaking.
4. To help children understand the world in which we live, and the interdependence of nations.
5. To help children to appreciate how the nation earns and maintains its standard of living and properly to esteem the essential role of industry and commerce in the process.
6. To provide a basis of mathematical, technical and scientific knowledge, enabling boys and girls to learn the essential skills needed in a fast-changing world of work.
7. To teach children about human achievment and aspirations in the arts and sciences, in religion, and in the search for a more just social order.
8. To encourage and foster the development of the children whose social or environmental disadvantages cripple their capacity to learn, if necessary by making additional resources available to them.

CONTACT WITH COLLEAGUES AND CHILDREN

None of these aims can be accomplished by shutting ourselves in our offices. The head has to be out and about. Although the managerial aspect of our job is becoming more time-consuming, we must not allow this to mean we lose touch with the children and the staff for whom we are responsible. We must always be a living part of the school, not a dead-weight.

In order to help the many different people we meet in the course of a day we should develop as many social and counselling skills as we can. Any experienced head will know that many people turn to us just because we are there. A large number of the problems brought to us by parents have little to do with their children, and only obliquely concern the school. We live in a society in which almost a quarter of all marriages split up. This means that we shall be responsible for some children who look to our schools to provide some stability and sense in their lives. No matter how tired or inadequate we may sometimes feel, we must always show a genuine interest in everything to do with our school.

QUALITIES

Being a head is fascinating work, but no one claims that it is easy, nor has it ever been so. The really effective headteachers seem to have been workaholics from a very early age. In *The Life and Correspondence of Thomas Arnold* by A. P. Stanley (12th ed, 1881), a colleague of the famous headmaster of Rugby wrote of him while he was still an assistant master:

> In the details of daily business, the quantity of time that he devoted to his pupils was very remarkable. Lessons began at seven, and with the interval of breakfast, lasted till nearly three; then he would walk with his pupils and dine at half-past five. At seven he usually had some lesson on hand; and it was only when we were all gathered up in the drawing room after tea, amidst young men on all sides of him, that he would commence work for himself in writing his sermons or Roman History.

In his survey of the qualities of the successful modern primary-school headteacher, *The Managerial Work of Primary School Headteachers* (Sheffield City Polytechnic, 1986), Alan Coulson also included energy, efficiency and effective management of time, but added self-confidence, an awareness of the sources of power and the ability to seek them out, and a sense of detachment to avoid being swamped by the responsibilities of the work.

Summary

In order to do the job properly, the modern primary school head teacher has to adopt and refine a style of leadership which will cope effectively with the many varied duties and responsibilities attached to the work. Some aspects of the style will come naturally, others will have to be acquired – perhaps painfully! This is a list of the possible qualities needed for the modern primary school headteacher:

- Perspicacity to assess the needs of the school
- Vision to formulate the aims of the school

- Flair to provide the ethos of the school
- Efficiency to organise the life of the school
- Tact and charm to manage the staff of the school
- Professionalism to appraise and train the teachers of the school
- Academic standards to help develop the curriculum of the school
- Commitment to learning to set an example to the school
- Firmness to carry out the necessary unpopular policies in the school
- Wisdom to counsel the parents and other visitors to the school
- Patience to teach the children of the school
- Attention to detail to manage the finances of the school
- Flamboyance to publicise the school
- Humility to learn to keep up-to-date for the good of the school
- Integrity to gain respect of the governors of the school
- Energy to keep in touch with everything going on in the school
- Idealism to expect the best from the school
- Teaching ability to hold your own with the staff of the school
- Health to withstand the pressures arising from running the school
- Humour and a sense of proportion to realise that no headteacher is going to possess all these qualities in sufficient quantities.

Self-assessment

The first task of any headteacher in this new era of education, whether a newcomer or more experienced, is to assess her needs. Which skills, concepts and attitudes do we need to improve in order to do our work properly? Are we too firmly locked into our administration? Could we be devoting too much attention to the theory of education at the expense of its practice? Is it possible that we are spending too much time in the classroom? Are we so intent on selling ourselves and our schools that there is too much show and too little substance? Which of the main leadership qualities do we need to work on to justify any claim that we are headteachers of 'whole' schools, strong in all areas? Are we neglecting any of the major duties of our employment? How do we really appear to our colleagues on the staff? Do they regard us as autocrats, benevolent despots, chums, or the first among equals in a professional team?

Try to take a little time to assess yourself and work out how you are doing professionally. It should pay dividends in the long run!

> **QUESTIONNAIRE**
>
> *Types of headship*
> Which of the four types of headship listed on pages 12 and 13 do you think you tend towards? Bearing in mind the constraints placed upon you by your personality, beliefs and experience, take a piece of paper and jot down an outline, realistic plan which could help you to include appropriate aspects of the three other leadership styles in your overall strategy for the school.
>
> *Attitude of headteacher towards staff*
> Which of the four attitudes listed on page 14 do you think that you take, on the whole, towards your colleagues? If you do not already incline towards the fourth type, try to draw up a blueprint which will help you steer your responses towards more open leadership, staff co-operation and full participation.
>
> *Duties of headteachers*
> Go through the list of the main conditions on pages 14 and 15. Jot down those aspects which you have not yet covered in your planning. Opposite each one write down how you intend to cope with it.
>
> *Aims of school*
> Study the list of suggested aims for a primary school on page 17. Draw up your own list for your school, bearing in mind its specific needs and the demands of the national curriculum.
>
> *Qualities needed*
> Study the qualities listed on pages 18 and 19. How many of them would you say you possessed? If you can lay claim to ten or more, you are doing well! Write down those qualities you feel you do not yet possess. Keep this list in a drawer and refer to it from time to time to see if you are making progress in these areas.

Don't feel depressed if you feel that you are being found wanting in some aspects of the headteacher's job. Welcome to the club! Just bear in mind that none of us can achieve an overnight transition to the ranks of superheads, if in fact there are any. The way to becoming an effective headteacher consists of mastering many details. These details will be described in the following chapters.

Chapter 2
School Organisation

Organisation is the framework upon which the structure of the whole school is built. It is also the most visible part of our leadership strategy and the one by which we are most likely to be judged, both by colleagues and onlookers. It does not matter how well-meaning our intentions, how sensitive our pastoral care, nor how erudite our curriculum development if our administration causes visitors to categorise our establishment as a second-rate school.

The overall aim of our organisation should be to enable the school to achieve its aims as fully as possible. It goes without saying that in any form of administration concerned with human beings and not robots, our organisation should be sufficiently flexible to take into account changing circumstances, staff losses and fluctuations in class numbers.

Organisation of time

No teacher's value to a school or influence upon the children may be measured by the number of hours worked, and different schools will organise their time in different ways. However, there are a number of obligations upon us as to the number of hours to be worked in the course of a school year, and it is the responsibility of the headteacher to organise this.

DIRECTED TIME
All teachers are expected to work 1265 hours a year, spread over 195 days. 190 of these will be teaching days and five will be non-teaching days devoted to in-service training. It is possible that there may be extra training days added to the school year.

This period of 1265 hours is known as 'directed time'. What happens during these hours should be decided by the headteacher in consultation with the teachers.

Teachers are also expected to work outside the 1265 hours, in what is known as 'non-directed time'. Individual teachers are left to decide for themselves how much time is needed in order to perform their duties efficiently. The main activities to be included in this time are:

- preparation of lessons
- marking of children's work
- writing reports
- school visits and journeys (outside normal school hours)
- courses (outside normal school hours).

As we all know, there is a third element to be added to the first two. This is the 'self-directed time', the countless hours which the vast majority of primary-school teachers spend on out-of-school activities such as clubs, sport and pastoral work.

Working out suitable hours

Headteachers in conjunction with their colleagues will have to work out which allocation of time best suits the needs of each individual school. It is advisable first to decide the hours of the working day, and then see how much time is left for the remainder of the directed time. An outline should be drawn up at a staff meeting.

There are a number of possible permutations, but in the average primary school most activities seem to take place between 8.30 am and 4.00 pm, although many heads and teachers arrive much earlier and leave a good deal later. This could be a possible outline of directed-time day:

8.40 am – 9.00 am	Non child-contact time	20 minutes
9.00 am – 12 noon	Teaching	3 hours
12 noon – 12.05 pm	Non child-contact time	5 minutes
	LUNCH	
12.55 pm – 1.00 pm	Non child-contact time	5 minutes
1.00 pm – 3.30 pm	Teaching	2 hours 30 minutes
3.15 pm – 3.50 pm	Non child-contact time	20 minutes
	Daily Total: 6 hours 20 minutes	

A day of this length, when multiplied by 190 working days, would give a total of 1178 hours, to be set against the total directed time of 1265 hours, leaving another 87 hours to be filled.

COMPOSITION OF SCHOOL DAY

The school day could differ in a number of respects from the example given. The headteacher, after due consultation, must allocate the hours in the best interests of the school.

In the case given, teachers would be expected to arrive at 8.40 am and be in their classrooms to meet parents and generally prepare for the school day. From 9.00 am until midday they would be teaching, with a break some time in the morning. Five minutes would be allocated at noon for clearing up and making sure that the children had gone home or were at the midday school meal.

The teachers would be expected to be back in their classrooms five minutes before the start of afternoon school at 1.00 pm. They would then teach until 3.30 pm, with a break for playtime. After the children had left for home, the teachers would have a minimum period of another 20 minutes in the school to meet parents, discuss matters with colleagues, conduct rehearsals for plays and concerts, and so on.

Technically speaking, the teachers should not spend any of their directed time in the periods before and after school in any non-directed time activities like marking or writing reports. However, if there were no other calls upon the teachers' time it would be a particularly meticulous head who would insist on colleagues abstaining from these duties.

The whole matter of directed and non-directed time is one for the head to approach with tact and care. Most teachers spend much more than the allocated hours on their work. A head who insisted on too rigorous a compliance with the timetable might find that her colleagues were beginning to work the minimum specified time, instead of their former much more generous self-allocated hours. On the other hand, of course, we are now spared the unpleasant spectacle of the class teacher who regularly sped from the classroom at hometime and reached the school gates ahead of even the fleetest children, or who practically rode her bicycle into the classroom at the stroke of 9.00 am.

DIRECTED TIME ACROSS THE YEAR

It is important to get the timing of the school day right and to make sure that it is understood by everyone. It is imperative that all teachers on the staff understand which of their duties come within the ambit of their directed hours. Should there be any conflict of views over this, there is always the chance that a head might find a teacher coming to her at the end of the Easter term with the news that her 1265 hours had already been worked out, with a term still to go.

The composition of those directed hours outside the school day will have to be determined by the particular needs and aims of the school, but there are some activities which are common to most schools and which all heads will want to include in their timetables. A possible outline of total directed time might be:

Working school day: 6 hours 20 minutes x 190 days	1178 hours
Out-of-school activities (concerts, fairs, discos, barbecues, parties, etc.)	20 hours
3 termly open evenings at 3 hours each	7½ hours
6 staff meetings at 1½ hours each	6 hours
Teacher appraisal	8 hours
5 non-teaching days at 6 hours each	30 hours
Leaving for contingencies and emergencies	15 hours
Total:	1265 hours

Headteachers should take care to leave at least 15 hours a year free to cover any unforeseen circumstances. It is quite impossible at the start of a school year to anticipate all the events which will have to be covered by the staff.

FLEXIBILITY

There is absolutely no need for the headteacher officiously to insist on every one of the 1265 hours being used up in the course of a school year. If everything is obviously going well, targets are being met, parents and governors are happy with the results, and the staff and children seem happy, busy and fulfilled, then the wise head should be satisfied that the spirit of the directed-time concept is being observed.

In the case of hours worked, the head will do well to remember that the self-motivated teachers on the staff will be spending a great deal of time on both non-directed and self-directed activities of which she may know little.

Organisation of the timetable

Once the basic teaching hours have been agreed, the head and staff can work out the timetable for the school. Much will depend upon the basic organisation of the school, and whether the approach is orthodox, open-learning, family grouped, or combinations of all these. With the emphasis on the core and foundation subjects in the national curriculum, no matter how the children are taught, a certain amount of time will have to be given to each of the designated subjects. It is only sensible to have this down in writing.

SCHOOL ORGANISATION

One of the most common methods is for the headteacher to draw up an outline timetable, based on the agreed school hours. After discussion with the teachers, decide upon the time to be spent on individual aspects of the curriculum. The basic components of the timetable will be:

Core subjects:	English, maths, science
Foundation subjects:	History, geography, PE, design and technology, music, art
Also:	Religious education
Others:	Registration, assembly, playtime, miscellaneous subjects: domestic science, health education, computer studies, etc.

If, for example, it was decided to devote approximately 75% of the timetable to the core and foundation subjects and religious education, and 25% to the other components of the school day, then the teachers could be asked to draw up their own provisional timetables, based on this assumption. The possible division of teaching time for a class might be:

School day: 9.00 am – 12 noon   1.00 pm – 3.30 pm =
            5 hours 30 minutes a day, or 27 hours 30 minutes a week

Foundation and core subjects and RE:	19 hours 40 minutes a week	(75% of total time)
Others:	6 hours 30 minutes a week	(25% of total time)

Core and foundation subjects and RE could be divided as follows:

English:	5 hours	Maths:	5 hours	Science:	2½ hours
Technology:	1 hour	History:	1 hour	Geography:	1 hour
RE:	1 hour	Music:	1 hour	Art:	1 hour
PE and games:	1 hour 10 minutes				

Total: 19 hours 40 minutes

The remaining time might then be divided into:

Assembly:	20 minutes × 5	=	1 hour 40 minutes a week
Registration:	10 minutes × 5	=	50 minutes a week
Playtime:	30 minutes × 5	=	2 hours 30 minutes a week
Miscellaneous subjects:			2 hours 50 minutes a week

THE MODERN PRIMARY SCHOOL HEADTEACHER

Example 2.1

	Morning					Afternoon			
	9.00	9.20	10.30	10.50	12.00	1.00	2.30	2.40	3.15
Monday	Assembly								
Tuesday	Assembly				TV Science				
Wednesday	Assembly		Hall PE			Field Games			
Thursday	Assembly				TV Music				
Friday	Assembly		Hall PE						

Example 2.2

	9.00	9.20	9.30	10.30	10.50	12.00	1.00	2.00	2.30	2.40	3.15
Monday	Assembly	Registration		Maths		English	History	Geography		Non-assigned*	
Tuesday	Assembly	Registration		PE Maths		TV Maths English	Technology	RE		RE	
Wednesday	Assembly	Registration		Maths		English	Games	Geography		Non-assigned*	
Thursday	Assembly	Registration		Maths		English Science	Science			Non-assigned*	
Friday	Assembly	Registration		PE Maths		English	Art	Music		Music	

*Non-assigned = finishing other work, or miscellaneous subjects, eg domestic science, health education

SCHOOL ORGANISATION

When the time to be devoted to each subject has been agreed upon, the head should draw up an outline timetable, based on the agreed school hours. The times already agreed for each class to take activities in the hall, television room, etc, should be filled in on the timetable blank by the headteacher as in Example 2.1.

These blank timetables should be distributed among the staff in the term before the teachers take their new classes, so that they have plenty of time to prepare the ratio of core, foundation and miscellaneous subjects. They should be encouraged to swap among themselves timetabled periods for the use of the hall, television room and resources centre, as long as these alterations are recorded on the timetables when they are returned to the head. It should be accepted that no teacher is going to be able to timetable subjects to the minute, especially as a number of them will almost certainly be amalgamated in project activities. The completed timetable might look like Example 2.2.

Each completed returned timetable should be checked with the teacher concerned in order to ensure that there is a proper balance in the curriculum of the class. These timetables should then be filed, pinned to the office wall, or transferred to the memory bank of a computer, so that both the head and the teachers have a permanent record of the school's curriculum in action.

Organisation of the headteacher's time

One of the basic tenets of any headteacher's life is that we shall rarely be where we should be. There will be too many calls upon our time and too many unexpected events for us to be able to timetable ourselves too rigidly. We will be that little bit more effective as leaders if we allow ourselves time and space in which to think and respond to crises.

Before we start planning our timetable, it may help if we try to work out how we are already using our available time. The results may be surprising!

The easiest way to do this is by keeping a brief outline diary over a period of time and then analysing the results. Filling in a log in this way may not satisfy the statisticians but should provide a rough-and-ready blueprint of how time is being spent. The log might work like this!

Date: Monday, 5th January

Time	Activity	Time spent
am		
8.30– 8.40	Arrive at school, open mail	10 minutes
8.40– 9.00	Informal talks with deputy and staff	20 minutes

9.00– 9.20	Take assembly	20 minutes
9.20– 9.45	Working with secretary on forms	25 minutes
9.45–10.15	Teaching class 6	30 minutes
10.15–10.30	Examining school with caretaker	15 minutes
10.30–10.50	Playtime, talking to teachers in staffroom	20 minutes
10.50–11.30	Teaching class 4	40 minutes
11.30–12.00	Meeting parent	30 minutes

After filling in worksheets for six or seven separate days spread over a month, the headteacher should study them and work out a number of main headings under which the time spent may be entered and examined. The breakdown of headteacher's time could be:

Average working hours: 8.30 am – 4.30 pm = 8 hours
plus: $1\frac{1}{2}$ hours working at home each evening
Total: $9\frac{1}{2}$ hours $\times$ 5 = $47\frac{1}{2}$ hours

General administration at home: $7\frac{1}{2}$ hours ⎫
General administration at school: 8 hours ⎬ 31% of total

Meeting parents and other visitors: 8 hours   15% of total
Teaching: 8 hours   15% of total

It is not possible to recommend any ideal allocation of time for heads, because circumstances will alter so much from school to school. We should examine the breakdown of our workload to see if there are any glaring anomalies. Is too much time being spent on administration? Is too much or too little time being spent in the classroom? Are there jobs which could profitably be delegated to others?

When we have worked out the sort of timetable allocation which seems best to meet our needs, we should use this to establish a system of priorities. We should ask ourselves what needs to be done next? It will help if we plan a series of actions which will enable each objective to be reached. For example,

Area: Building and Plant

Immediate aim:	Mid-term aim:	Long-term aim:
Raise funds to convert old kitchen into library	Hire contractors to do work	Have gala opening by celebrity

ADAPTING

In all our planning we should be prepared to adapt to changing circumstances. This was brought home to me when once I decided to make myself more accessible to parents before morning school each

SCHOOL ORGANISATION

day. For a number of mornings at about 8.30 am, when the first parents began to assemble at the school gate, I would head in their direction in a manner likened by irreverent colleagues to a minor member of royalty going walkabout.

The results were far from encouraging. At first the assembled parents seemed mildly affronted that I should seek to interrupt their early-morning get-togethers. However, once they had grown accustomed to my presence, the reverse occurred. I was so inundated with requests and petty complaints that I could not cope with the sheer volume of traffic, but there was still a need to make myself available in this informal manner. I knew that most parents would be reluctant to come to my office. Instead, I began to take up my station in the school hall for half an hour every morning. When staff and parents realised what was happening, many of them took advantage of my presence there to come and see me casually. In this way, I was able to maintain contact with everyone who really wished to see me informally, but by retreating into the school I discouraged any professional complainers who could not be bothered to walk across the playground into the school hall.

The school day

Changes are taking place in the time and composition of the school day. Governors now have the right to decide the hours which will be taught and the time that the school will start in the morning, and the time that it will stop work. It is generally agreed that $21\frac{1}{2}$ hours a week for infants and 23 hours a week for juniors should provide sufficient time for schools to do their work, but that these hours should not include the time spent on assemblies or playtimes. The minimum number of sessions is 380 in each academic year. The academic year is deemed to start from 1st September, for a period of twelve months, unless the school year begins in August, in which case August 1st is the starting date.

Attendance of children

Children are expected to attend school regularly once they have been admitted. Registers must be marked at the start of each morning and afternoon session, using ink. These registers must be kept for at least three years after they were last used.

Any child who has been absent for a continuous period of not less than two weeks should be reported to the LEA, unless a medical certificate has been sent to the school. Permission may be given for a child to be taken on holiday during term time for a period of not longer than two weeks.

Admission numbers

The 1988 Education Reform Act states that each school must have a standard number based on the physical capacity of the building. This number will be based on the total number of children admitted to the school in the whole of the year 1979–80, or the number of pupils taken in the year prior to the enactment of the legislation of the Act, whichever is the higher.

All schools are expected to admit as many pupils as they can comfortably accommodate, but they are not expected to expand beyond their physical capacity. This means that popular schools may have to turn pupils away.

Data Protection Act, 1984

If a school uses a computer to help with the organisation of facts about staff, parents or any other members of the community, the provisions of the *Data Protection Act* will apply. There are eight basic principles which must be applied to all living persons:

1. The information contained in personal data shall be obtained and processed fairly and lawfully.
2. Personal data shall be held only for one or more specified and lawful purposes.
3. Personal data held shall not be used or disclosed for any unlawful purposes.
4. Personal data stored in this fashion shall be adequate, relevant and not excessive.
5. Personal data shall be accurate and, where necessary, kept up-to-date.
6. Personal data shall not be kept longer than is necessary for that purpose.
7. An individual shall be entitled, at reasonable intervals and without undue delay or expense to be informed of any data held, to have access to it and, where appropriate, to have such data corrected or erased.
8. Appropriate security measures shall be taken against unauthorised access to, alteration or destruction of personal data, or its loss.

SCHOOL ORGANISATION

Summary

It is the duty of the headteacher to organise the working hours of her staff and arrange the school timetable so that the national curriculum is covered effectively. The headteacher must also arrange her own working time so that priority is given to the most important tasks. The head should consider the major objectives of the modern primary school headteacher and decide how much attention needs to be paid to each one.

Self-assessment

We shall probably be judged more by the standard of our organisation within our schools than anything else. We should make sure that the administration of time and resources within our schools is efficient. Is the best use being made of teachers' time? Do our colleagues know what they should be doing and when they should be doing it? Have we got our priorities right? Is the organisation of learning within our schools properly structured? Are all children in the school receiving an equal share of the skills and resources available?

QUESTIONNAIRE
Directed time
Revise the 1265 directed hours for the staff of your school in the light of the suggestions on page 24. Leave yourself with at least 15 hours, and preferably more, to cover any emergencies.

Timetables
Examine the timetables of the classes in your school and compare them with the suggested distribution of time to subjects in the examples given on page 25. Do you feel that you should make any alterations, or are you satisfied with your existing arrangements?

Headteacher's time
In the light of the information provided on page 28, keep a log of your own activities over a period of six or seven days. Do you think that you should reorganise your duties as a result of this?

Chapter 3
The Children

The relationship between the primary-school headteacher and the pupils of the school can be a complicated one. Getting it right is central to the whole business of running an effective school, and presents a problem to all but the most charismatic and dynamic of heads.

Most class teachers worth their salt can build up a solid, affectionate and enduring bond with the majority of the thirty or so children in their charge. It is one of the first things that they will try to do. A head, who may be responsible for up to six hundred boys and girls, will find it less easy to establish such a helpful connection.

The headteacher's image

On the whole, as a body, headteachers have never enjoyed a good press. No matter how hard we may work at our image-building as individuals, most children are going to have unflattering and off-putting conceptions of us. In the comics and books that they read we are usually depicted as ogres, fanatics or incompetents. Our teaching colleagues, members of the community and the parents of the children use us as a form of ultimate deterrent.

It is then hardly surprising if the children in our charge regard our arrival with the same suspicion as they would a visit from a travelling dentist. Some headteachers can soar above such a reception, and by sheer force of personality develop strong links with their pupils. Most of us, however, have to roll up our sleeves and earn the respect if not the adulation of the children.

GETTING ABOUT THE SCHOOL
We must be more than remote, meaningless figures glimpsed dimly in the hall, doling out information, admonitions and threats. Obviously

the boys and girls will relate much more easily to the class teachers, with whom they spend so much of each school day, and this is as it should be. It is important that they should also learn to appreciate the role of the headteacher as an integral part of the daily life of the school.

This may be accomplished in a number of ways. Above all, the headteacher should constantly be on the move in an unhurried manner about the school. In management jargon she should maintain a high visibility profile. This means we should organise our daily timetable so that the children become accustomed to seeing us working about the school in a controlled and orderly fashion for an appreciable period most days of the week. Were it not for the fact that we need somewhere for confidential meetings, I would board up all headteachers' offices, or use them for a sensible purpose like storerooms.

As it is, we should take as much of our paperwork as possible with us and do it wherever there is a suitable surface in the school. We could use the library, dining-hall and anywhere else where the children will see us as they pass from one classroom to another, and perhaps stop to talk to us from time to time. While we should certainly take a full part of the teaching load on our shoulders, we should take care not to immerse ourselves for too long in one classroom, where the rest of the school will forget us.

As far as possible we should restrict our time in the office to certain definite periods, because once we get in there are so many problems demanding immediate solutions, we may never get out again. Instead, it is easy enough at the beginning of a week to try to schedule all meetings with visitors for, say, the last hour of each afternoon. In the same way, at the beginning of each day we could make a note of all the telephone calls we have to make, and arrange to conduct these in one block at a particular time of the day.

Of course there will always be unexpected visitors and urgent calls to make, but most of the administrative work of this nature can be arranged for a stated time each day, so that staff and parents know when it is easiest to contact us.

It will not be long before the children grow accustomed to seeing us about the school, working in open spaces, teaching groups or classes in the hall, helping individual teachers in the classrooms, and generally assisting in the daily life of the school.

KNOW THE CHILDREN
While we are ensuring that the children are aware of us as an integral part of the life of the school, it should also be one of our duties to know

the name and something about each of the boys and girls for whom we are responsible.

In large primary schools this will be difficult, but it will be worth the time and effort involved. If there are many names and faces to be recognised, it might help to have small snapshots of the children and attach them to their record cards. Most schools hire professional photographers to come in once a year to take portraits of the children to sell to the parents in aid of the school fund. These photographers will usually provide miniature versions of the photographs, either free of charge or for a nominal sum. Failing this, the children could be asked to bring their own photographs with them from home. If the headteacher examines the likeness every time a card is consulted it should not be long before she can put a name to most of the children.

Children, like adults, appreciate it if we address them by their first names and let them see by some question or remark that we know something about them. They may seem to take such encounters for granted, that is the way of most children, but such brief individual moments of contact really mean a great deal to young boys and girls and help contribute to a warm and caring ethos within the school.

GAINING THE RESPECT OF THE CHILDREN

The next step is to gain the respect of the children. Our own individual personalities will play a large part in this, but respect is also gained by working hard and being good at what we do. Children do not take long to grasp when the headteacher is really making an effort and when she is just going through the motions.

It is up to each of us to prepare our lessons and make them interesting and relevant. This is easy enough when we are timetabled to take a class regularly. However, when colleagues are absent we shall often be pitch-forked into a class to take over at short notice. More experienced heads will have built up a store of materials to draw upon in such cases. Others may not be so fortunate. The head whose response to taking over a class unexpectedly is to growl 'Turn to page five in your Maths books', will soon lose the respect of children and colleagues alike. It is far better to have our own stock of emergency lessons.

Children will take their attitudes from the teachers and other members of the staff. If our colleagues obviously accept our position as a head, then so should the children. It is a professional responsibility of teachers to accept the status of the head, just as it is the headteacher's duty to be dignified and responsible, without being cold or remote. If our relationships with the staff are good, then contact with the children will be that much better and easier.

We should always be polite and considerate in our dealings with everyone in the school, adults and children alike, and insist, quietly but firmly, on being treated by others in the same manner. Never criticise any member of the staff in the presence of colleagues or children.

Establishing one's position in this way should not as a rule present any great problems. Even in today's egalitarian society most adults and children accept the need for leadership within a society or institution. The form that it takes may be a subdued and thoughtful application of influence rather than any rigid exercise of authority, but it should be just as effective. If we make it our business to know the children and the staff, and perform our tasks in a friendly and straightforward manner, then everyone in the school should be only too willing to acknowledge our position and co-operate to the full.

We should also make sure that our efforts at communication are a two-way affair. We should take time to ensure that the children understand us and are responding to our efforts. In my own case it came as a salutary experience to discover that despite my endeavours to appear a low-key operator, at least one eight year-old girl in the school still regarded me as a fairly typical adult grouch!

>
> Quiet
>
> It
> is nice
> God tells us
> to be quiet.
> You cannot see it.
> Our Head says it
> all the time
> to us
> all.
>
> Lynne Waddell

CONSISTENCY
As a part of our general image in the eyes of the boys and girls, we must earn and deserve a reputation for being fair and consistent. This is best accomplished by avoiding hasty decisions. No matter how busy we are we should always take time to consider matters concerning the pupils, to hear all sides of every case, and wherever possible to give reasons for any decisions we make. If the headteacher cannot be considered well-balanced and reliable, then to whom can the children look for stability?

The headteacher as a teacher

There is considerable pressure upon us all to spend a great deal of our time teaching in the classroom. Classes may be large, teachers are overworked; there is always a demand for an extra pair of hands up at the sharp end.

Some heads, of course, have little option. In 1988, there were eight thousand primary schools with full-time teaching heads and many others whose headteachers had to take on heavy teaching loads. For those of us with schools of more than 200 children there is usually more choice as to how we allocate our time, but it would be a mistake and almost a physical impossibility to spend too much time teaching.

Most heads enjoy teaching, they may have been promoted because they were outstanding class teachers. It is essential that we put aside some time to keep our hands in as teachers and to get to know the pupils – it can be very pleasant to get away from our problems and immerse ourselves in the detail of classroom teaching. However, it was not what we were appointed for. We have more important things to do!

Nevertheless, we should spend regular periods of time helping with teaching. In order to keep up-to-date with what is going on in the classrooms, we should divide our child-contact time into three main sections.

TIMETABLING OURSELVES
The first and perhaps the most important of these is to timetable ourselves so that we spend time regularly in the classrooms helping the teachers. If it is at all possible we should try to assist a different teacher for an hour each day. In a normal seven-class 5–11 school, this will mean that we should get around to each classroom for an hour every seven days. In larger schools it will take longer, but even in the biggest we should be able to spend an hour working with each teacher and class every two or three weeks.

This period should be spent as a combined teacher and ancillary helper, assisting the teacher in any way that she asks. The work could consist of taking a group, or going from table to table helping the children at whatever tasks they may be undertaking – listening to children read, working with the slow learners, and generally giving a hand.

In this way the headteacher makes contact with the children on a regular basis and is able to judge their progress and the general ambience of a class. At the same time it is possible to observe each teacher at work and become aware of any undue pressures or stress.

THE CHILDREN

TEACHING
The second way in which the headteacher can make regular contact with the children is by teaching them without the class teacher. There will be times when it will be sensible to take a single class – to allow the deputy head time for administrative chores, or while teachers are away ill or on courses for example.

As a rule, however, it will save time and be more convenient if the headteacher arranges to take groups of classes at a time. This may best be done by combining classes for television programmes, or lessons based on videos or filmstrips. Other areas in which the head could take groups of classes include assemblies, singing, and so on. Perhaps 45 minutes a day could be allocated to this teaching.

ASSEMBLIES
The third method of making use of the headteacher is the school assembly. This could consist of taking the whole school together, or groups of classes. The assembly should really be the centre-piece of the headteachers' public image and should be approached with great care and attention.

All pupils should take part in a daily act of collective worship. This may take place at any time and could be a single act for the whole school, or separate acts for different school groups. Most of these acts of worship must be of a broadly Christian character and appropriate to the family backgrounds, ages and aptitudes of the pupils.

Experience suggests that if time and space allow, most assemblies should include the whole school. These are among the few occasions when it is possible for all the children to share in an activity. If there is not enough room for this, then separate acts of worship should be arranged.

The assembly is also important because it is a function at which the performance of the headteacher may be seen and judged by children and staff alike. In a brief twenty-minute period, the head has a chance to display organisational ability, teaching qualities and even a brand of showmanship.

If our assemblies are reasonably original, interesting and inspirational, then the children and teachers should be able to accept that we are in command of an important part of the headteacher's craft, that of being able to put over ourselves and our message.

With the exercise of a little ingenuity, we can even use our assemblies as the centre of the school's religious education syllabus.

This may be accomplished by the headteacher taking the first three of the week's services, perhaps using the BBC's religious service for schools by radio on the fourth day for a little added variety, and then asking the teachers and children to present the final service of the week, or to present such a service every other week.

At the beginning of the school year, a list of forty or so themes for assemblies could be drawn up, one for each week of the year. The themes chosen could include the main celebrations on the Christian calendar – Christmas, Easter, and so on; the lives of great Christians, and common themes capable of being illustrated through the Christian faith and other beliefs – Truth, Beauty, Friendship, etc, and the main celebrations on other religious faiths' calendars.

The first three assemblies of the week, conducted by the headteacher, should be based on the agreed project for the week. For example, if the theme were *Signs*, then all three of the head's assemblies should be concerned with that topic and might be outlined as follows:

Theme:	Signs
Introductory music:	'The Ink is Black' – The Spinners
Display:	Exhibition of children's written work
Children's activity:	Reading selected poems on religious or moral theme to illustrate beauty of language
Hymn:	Based on theme of signs or symbols, e.g. 'Give me Oil in my Lamp', 'The Old Rugged Cross', etc.
Head's talk:	Explain symbolism in chosen hymn. Discuss significance of signs and symbols. Ask children to look for signs around them in preparation for their own assembly later in week.
Reading:	Story of Emperor Constantine seeing sign of cross in sky, and becoming a Christian
Closing hymn:	Based on sign or symbol, e.g. 'The Lord's my Shepherd', 'When a Knight Won his Spurs', etc.

Each class in the school would be asked to take one aspect of the theme as the basis for its Religious Education syllabus for the week, and to prepare a short item for inclusion in the children's assembly, to take place weekly or fortnightly. This means of course, that the headteacher will have discussed the theme with the class teachers, so that they have had time to get ready for it with their children.

In order to make the children's contribution even more effective, the headteacher, in consultation with the staff, could draw up specific

THE CHILDREN

suggestions for each age group within the school. These are some possibilities:

Religious theme: 'Talking to God'	Religious theme: 'God loves us'	Religious theme 'Signs of God'
Assembly approach 'Signs we make'	Assembly approach 'Signs we make'	Assembly approach 'Signs we see'
Praying Kneeling	Shaking hands	Church signs
Saluting Waving	Embracing	Road signs Clothes that are signs: turbans
1st year infants	*2nd year infants*	*3rd year infants*

Theme: Signs

1st year juniors	*2nd year juniors*	*3rd year juniors*	*4th year juniors*
Religious theme 'Belonging to God'	Religious theme 'God cares'	Religious theme 'Celebrating God'	Religious theme 'Seeing God around us'
Assembly approach 'Belonging signs'	Assembly approach 'Caring signs'	Assembly approach 'Celebration signs'	Assembly approach 'Signs of faith'
Badges Flags	Hospital signs	Festivals of Church	Christian signs
Church buildings	Church Army signs		
Colour of skin: belonging to different groups	Red Cross and Red Crescent signs	Festivals of other churches: Hindu: Divali Islam: Ramadan Jewish: Passover	Signs of other faiths
		Birthdays Anniversaries	

Each teacher in turn could take the general heading suggested for her class as the basis for that week's religious education project, as well as the foundation of her assembly contribution. Each theme is broad enough for the teacher to treat it in any way she liked. A different teacher could be asked to contribute to assembly each week. The composition of a children's assembly on theme of 'Signs' might be:

1st Year Infants:	Say a prayer they have learnt
2nd Year Infants:	Demonstrate ways of shaking hands, e.g. before a boxing match, as a sign of friendship
3rd Year Infants:	Talk about church signs they have seen: cross, etc.
1st Year Juniors:	Demonstrate models of uniforms of faith: priest's surplice, Salvation Army hat, etc.
2nd Year Juniors:	Demonstrate and talk about caring signs they have discovered: hospital, hospice, etc.
3rd Year Juniors:	Mime a religious festival: harvest, etc.
4th Year Juniors:	Demonstrate sign of cross they have made and talk about signs of other faiths: crescent, star, etc.

Obviously such assemblies will have to be tailored to fit the size of the school concerned. Larger schools may need two children's assemblies a week to include all the contributions. Other schools may find it more practicable to have a children's assembly only once a month.

Not all heads will have the time or the inclination to base the school's RE syllabus on the morning assembly, but if they only use the time for the traditional prayer, hymn, story and notices, considerable care and attention should still go into the preparation. If our assemblies are poorly planned and delivered in a perfunctory manner it will not be long before we lose the respect of both children and colleagues, and rightly so.

Discipline

Discipline is an important part of the ethos of a school, but its ramifications spread far wider. A school is judged by its community more by the behaviour of its children than anything else. It will take a primary school years to build a reputation for its academic and sporting excellence; the bad manners of a few of its children spilling out into the streets can condemn a school in the mind of outsiders immediately.

The neighbourhood is entitled to expect a reasonable standard of conduct from the boys and girls attending its primary school. This should be and usually is the responsibility of the parents, in collaboration with the school. Where there is not a great deal of support from a minority of parents, then the school simply has to try to do all the work.

If the school is to do its job properly, the headteacher must see to it that nothing comes between the children and the full range of opportunities being offered them. Bad behaviour is a distraction. The headteacher must establish a method of dealing with disruptive acts. This will become the school's code of discipline.

The fewer rules a school has, the easier it will be to enforce them and the easier it will be for the children to remember them. Posting a long list of do's and don'ts on the corridor walls, or ranting at assembly will not as a rule make much impression on children.

It has been my experience that those schools which seem to run with a minimum amount of disturbance limit their philosophy of behaviour to the one word, respect. Below is one flow-chart which could with profit be displayed in the school's entrance hall and included in the brochure. It is easy to understand and gives the pupils the opportunity to think about the edict and in many cases impose their own self-discipline.

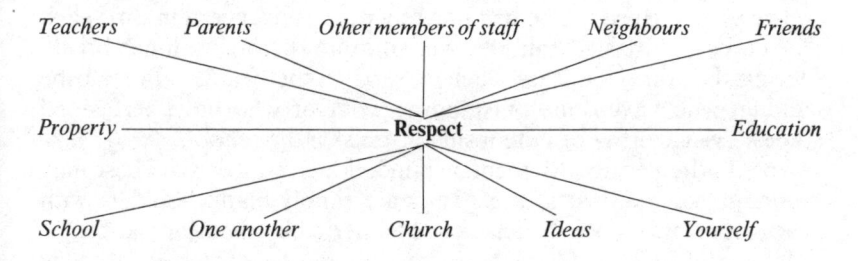

Teachers	*Parents*	*Other members of staff*	*Neighbours*	*Friends*
Property		**Respect**		*Education*
School	*One another*	*Church*	*Ideas*	*Yourself*

If the children are encouraged to understand that it is up to them to apply the principle of mutual respect in their dealings with others inside and outside the school, most of them will see the reason for such a code of conduct within a community, and the importance in a busy and crowded environment of a spirit of co-operation and give-and-take.

Once this basic philosophy has been established, it will still be necessary to spell it out in one or two cases. A few elementary ground rules should be established, so that all children know what is expected of them. These rules may change from school to school, depending on individual circumstances, type of building and playground, and so on. In all schools they should include an insistence on courtesy to adults, consideration for one another, and no fighting or bullying.

It is a truism, but one worth repeating, that if the children are kept busy and interested in the classroom and properly supervised outside it, the opportunities and perhaps even the desire for wrong-doing may be minimised.

Once the code of conduct and the basic rules of behaviour have been drawn up by the headteacher and staff, agreed by the governors and put into practice among the children, most schools will then institute a system of rewards and punishments in order to back up the school's code of discipline. There may be some heads who object to the goad-and-carrot system, but it can be an effective method of ensuring that the school runs smoothly, especially if the rewards outnumber the punishments.

REWARDS

The most effective system of rewards that I have seen in use in schools is the merit principle.

The main advantge of this system is that the selection of pupils for the merit roll remains in the hands of the teachers, who will know the children better than the head does, yet at the same time it enables the headteacher to meet children on a one-to-one basis regularly in an extremely pleasant context.

The system is easy. The head merely puts aside a certain time each day, perhaps the last 45 minutes every morning before the lunch break. During that time the class teachers send to the head's office all the children who have done outstanding work or who have performed some service worthy of note inside or outside the school.

The children bring with them their individual merit cards. These may be bought or made at school. They are simply blank booklets with spaces for stars or merit stickers to be inserted by the headteacher.

Each child shows his work in turn to the headteacher and they discuss it together. Such individual contact helps the head to get to know the children and assess the standards of work being produced in the classrooms. The children in turn have the experience of meeting the headteacher in a pleasant, friendly and non-threatening atmosphere.

When the work or action has been discussed with the head and the child has been praised, the child's name is written in the Merit Book. This can be a large diary or a notebook kept specially for the purpose. Then the merit sticker is put in the child's book or card. Once a week, all the children in the Merit Book could be called out in front of the others at assembly in order to display their work to the whole school.

Other forms of reward which the headteacher could use include constant praise of deserving children, mention of these children in the school newsletters, a list in the hall on which children are allowed to enter their own names after a good piece of work, letters home to parents, and examples of good work placed in the foyer or hall, with the names of the children responsible prominently displayed.

PUNISHMENTS

The subject of punishments is much more difficult. A great deal will depend upon the attitude of the individual head towards dealing with infractions of the school rules and disruptive children.

Under no circumstances must any form of corporal punishment be used – striking, shaking or pushing a child. The only circumstances in which a headteacher might be justified in handling a child would be in order to avert physical harm to that child, by separating two children who were fighting, or something of that sort. Even here the use of the voice should be tried first.

In most primary schools anyway, the use of corporal punishment has been on the decrease for years. In an ordered and caring environment, with teachers who know their pupils and are swift to notice and investigate any out-of-character actions, most children should be content to conform to the norm of behaviour.

It must be accepted that there will usually be a few boys and girls who, for various reasons, will break the rules. Often the cause may be boredom, an overcrowded playground, sheer thoughtlessness, or a desire to 'try it on'. In such cases the head should back up the class teacher concerned immediately. A warning, followed by a mild punishment such as being deprived of a playtime or a games lesson will often suffice.

Another response could be for the errant child to work in the headteacher's office under her supervision for a few periods. If nothing else this will give the teacher a rest! For more serious or frequent breaches of discipline, the head will need to examine the child's background, in case some form of counselling or other pastoral care is indicated. Has there been any change in the child's circumstances? Are there problems at home? Has the child fallen out with a friend of long-standing? Is he unwell? Is he being bullied?

In cases like these, the headteacher should have a quiet and confidential talk with the child in question. It may be that the boy or girl has been driven to some anti-social act merely to attract attention. If the reason for this cry for help can be ascertained, it is possible that some solution may be found.

In many cases, unfortunately, we will not be able to discover the cause of disruptive behaviour on the part of a naughty or aggressive child. It is then that we shall have to widen the scope of our enquiries and involve the parents and such supportive agencies as the Education Welfare Service, the Educational Psychologists' Department, the Child Guidance officers, and so on. The headteacher should make a point of developing a good working relationship with all these organisations.

Consulting parents
Asking parents to come in and talk about their children who are experiencing difficulties with their behaviour can be useful and productive. The head will have to judge the right moment to involve parents, but unless the case is a desperate one it is better to try to solve the problem within the school first.

However, parents have a right to know if their children are in trouble at school. Some headteachers are nervous about bringing in mothers and fathers, but most parents are only too well aware of their children's deficiencies, and will welcome a chance to collaborate with the school in an effort to do something about it. It can also be very effective when a child discovers that he can no longer play off the home against the school, and vice versa.

If, however, the parents refuse to come in to the school, or show little interest in helping their children, and if the educational psychologist has also drawn a blank with short-term responses, then the headteacher may have little choice left but to consider excluding from the school a child who persistently behaves badly.

Drawing up an exclusion policy for the school
The headteacher should draw up a general policy on exclusion and present it to the governors for their approval.

Only the headteacher has the right to exclude a child. There are three types of exclusion:

- *Temporary*: no more than five days in aggregate in one term without informing the governors and the LEA. If the exclusion is for more than five days, or would debar a child from taking a public examination, the LEA and governors must be informed.
- *Indefinite*: no fixed time for return of child.
- *Permanent*: expulsion of the child.

The headteacher should inform the parents at once of the decision to exclude the child, the length of the exclusion and the reason for it. Every effort should be made to contact the parents by telephone, and a following letter should be sent by first-class post, confirming the exclusion. The parents should be informed of their right to appeal against the exclusion to the governors or the LEA. The governors have the right to direct a headteacher to take the excluded child back into the school, and so does the LEA, but it must consult the governors first.

BULLYING

Bullying in a school usually takes place in the playground before and after school, at playtime and during the lunch break, if it takes place at all. It is a prime duty of the headteacher to make sure that the playground and school buildings are patrolled by responsible, vigilant adults whenever children are on the premises but not actually being taught in the classrooms.

The determination of the school not to tolerate any form of bullying should be stated in the school brochure, and both parents and children should be made aware that bullying is certainly one practice that may be punished by exclusion from the school.

Staff should be instructed to be on the alert for any signs of 'gangs' or of individual persecution in the playground, and asked to pay particular attention to any complaints from children that they are being picked on. All children should be encouraged to report any type of bullying by other boys and girls at once to an adult on the staff. In their book *Bullying in Schools* (Trentham Books, 1989), the authors, Delwyn Tattum and David Lane, emphasise the importance of a school policy 'aimed at consciously creating a caring, concerned, sympathetic ethos in which all members are respected'.

If there do seem to be signs of an outbreak of bullying, teachers and the head should act at once. All teachers on playground duty should get out to the area straight away at playtime, and lunchtime supervisors should be similarly briefed. Hideaways like toilets and spaces behind the bicycle shed where the bullies may lurk should be watched carefully.

When the bullies have been identified they should be seen by the headteacher. They should be left in no doubt about the serious view taken of their actions, and the punishment likely to befall them should be described in detail. It sometimes helps to allow a bully to be on the receiving end of rough treatment for a short while, for example by placing an aggressive boy in a school football match, closely supervised, where he will be with better players and could take a few knocks.

The children who have been bullied should also be seen by the head, or a sympathetic delegated teacher, and an effort made to build up their self-confidence and self-esteem.

In addition to the pastoral approach, the problem of bullying and aggression could also be dealt with in the academic curriculum, particularly in RE and assemblies, and discussions conducted in which the subject is aired.

It should be remembered, however, that children do fall out, and that some accusations of bullying will turn out to be unjustified. Even so, no complaints should go uninvestigated.

Truancy
The problems of children who habitually stay away from school have been studied by many academic researchers. The reasons for such absences are not difficult to find, but discovering a remedy for truancy is something else again.

Traditionally truants are either low achievers who see little purpose in coming to school, or anxious and sensitive children who are unable to cope with the problems of community life and try to escape, both literally and metaphorically, into worlds of their own.

Encouraging children with a school phobia to give school a chance is far from easy. In his book *Truancy* (University of London Press, 1968), M. J. Tyerman states that the single most powerful influence on school attendance is a sense of purpose in the school felt by parents, teachers and pupils alike.

The co-operation of parents in ensuring that children attend school regularly is essential. When that assistance is forthcoming children can usually be encouraged to come to school. If the parents are not interested, then the school can do relatively little.

Some schools encourage good attendance by handing out certificates for 100% records in the course of a year. I was put off that method for life when one child in the class I was teaching at the time came up to me and whispered, 'I would have got that certificate if my Mum hadn't died and I had to go to her funeral.'

The experience of the Education Welfare Service may be invaluable in the case of persistent absences from school on the part of a child, and every effort should be made to discover *why* the child is reluctant to come to school. When this is known, with the co-operation of parents, it may be possible to help the child overcome its problems, but it will be a long, slow haul.

A Code of Discipline
When a headteacher comes to draw up a code of discipline for a school, one of the most helpful documents to consult will be the HMI publication, *Education Observed 5; Good behaviour and discipline in schools* (DES, 1987). It points out that teachers in this country are

expected to accept responsibility for the social development of their pupils to a degree unknown in the rest of Europe, and goes on to highlight some principles necessary to maintain high standards of behaviour and discipline in schools.

1. Policies on behaviour must be explicit, provide guidelines for action, and be firmly and consistently applied.
2. A positive climate must be developed for the whole school, based on community of purpose, consistent practice and constant vigilance.
3. The range and application of rewards must outweigh those of sanctions.
4. The school's leadership must set a good example. Quality of relationships at all levels is where the ethos of the school is grounded. It must obtain not only between teachers and pupils, but among teachers and among pupils. All the school's activities must contribute to this climate.
5. The school must make use of the wider partnership with parents and the community.

In the same way, the school should publish its code of punishments and sanctions. The school brochure would be the best place for this, when it has been agreed with the governors. For example:

If a child misbehaves, the following are the steps that will be taken at the school:
1. The pupil will be rebuked by the class teacher.
2. The pupil will be rebuked by the headteacher.
3. The pupil will lose a playtime or games period.
4. The pupil will work in the headteacher's office for a specified time.
5. The parents will be sent for and asked to help.
6. The services of the educational psychologist or the Child Guidance Service will be enlisted.
7. The child will be excluded from the school.

Dealing with problems of discipline usually boils down to vigilant teachers backing up an explicit code of conduct and nipping any problems in the bud, dealing firmly with transgressors and counselling victims wisely and sympathetically, all with the firm and consistent backing of the headteacher.

CARE AND SUPERVISION
The headteacher is ultimately responsible for the care and welfare of

the children in her charge and should draw up detailed plans for the safety of the children on school premises before, during and after school hours. She should make sure that all members of the staff and the children are aware of the arrangements made for their care and protection. An outline policy for the surveillance of children could include the following points.

Make sure that parents and children know exactly what time the school starts and finishes. Children should be admitted into the official care of the school some fifteen minutes before the start of morning school. It is advisable to have teachers on duty in the playground as well as the classrooms for those fifteen minutes.

There must be a definite policy as to whether children are allowed into school indiscriminately before the start of morning school, or only when the weather is bad. If the children are admitted there must be a teacher on duty in the classroom or appropriate section of the school receiving them.

There should be a rota of teachers to supervise the children at playtime; these teachers should go out to the playground as soon as the children are released.

If the weather is too poor to allow the children to go outside at playtime, there should be teachers on duty in each classroom or section of the school housing the children.

There must be adequate arrangements for children to enter and leave the school premises in an orderly manner at hometime.

At lunchtime, the children on the premises must be supervised carefully by attendants or volunteers from the staff, issued with written instructions as to their duties. There is no compulsion for the head to remain on the premises at lunchtime, but if she leaves, a deputy or other teacher must be nominated to be in charge.

All dinner supervisors must be told that they have no authority to punish children for breaches of the rules, but should report children to a member of the teaching staff.

Under no circumstances should children be allowed off school premises during school hours, unless a letter requesting this is received from a parent or guardian. Even then the responsible adult should be requested to collect the child from school.

Children who remain behind after school to take part in extra-curricular activities are still the responsibility of the headteacher, who must ensure that there is adequate supervision of the children as long as they are on the school premises.

SAFETY OF PREMISES

As a matter of urgency, the headteacher should check the school premises regularly to ensure that nothing has occurred which could present a hazard to the safety of the children. Any holes in the playground, damaged equipment, etc. should be repaired immediately, and the children kept well away from any hazards or potential hazards until it is safe to approach them again. The school caretaker, if there is one, should have the immediate responsibility of reporting any likely dangers to the head.

RESPONSIBILITIES OF TEACHERS

During the school day the responsibility for the safety of the children will rest in the first instance upon the teachers, but the head is still in charge of ensuring that the standard of care and protection provided is adequate.

Teachers should see to it that the children are not left unsupervised in their classrooms, that any lessons like PE and craft and technology which involve the use of potentially dangerous equipment or apparatus are carefully prepared and monitored.

It is the responsibility of the head, who may delegate it to a named member of the staff, to ensure that there are properly stocked and replenished first-aid boxes situated in the school.

The headteacher is also responsible for the formulation of a policy to be followed in the case of fire, to organise a system of fire drills on a regular basis, and to ensure that fire-bells work and can be heard by all teachers and children. A suggested outline fire policy for a school might be:

- Any child seeing a fire in the school should tell the nearest grown-up.
- Any adult discovering a fire should at once give the alarm. This should be done by giving a long continuous ring on one of the fire-bells. These fire-bells are situated as follows ...
- The glass on the fire-bell should be broken with the hammer attached to it, or with a blow from a shoe or heavy object.
- Upon hearing the fire-bell, all teachers should at once assemble their children in an orderly and calm manner and take them out to their allocated places in the playground, unless these assembly-points are obviously dangerous, in which case the teacher should move the children to a safe spot.
- Before leaving the classroom the teacher should try to ensure that all doors and windows are shut, if time permits.
- Unless it lies in the path of a fire, teachers should take their children

out through the safety, i.e. fire-doors. These doors should always be unlocked as soon as the teacher enters the classroom in the morning.
- Teachers should take the registers with them into the playground and call the roll once they have assembled their children. Classes should be reported present to the Headteacher or Deputy.
- No child or adult will go back into the school without the express permission of the Head or Deputy.
- Ancillary staff will assemble in the playground when the bell rings, and report to the Headteacher.
- As soon as the bell rings, the Headteacher and Caretaker, if the latter is on duty, will explore the school to make sure that no child has been left behind. The Headteacher will have summoned the fire brigade as soon as the fire was reported to her.
- The Headteacher and Caretaker will take all practicable steps to put out the fire, but the first responsibility of all adults is the safety of the children.
- Fire extinguishers are situated ....

EDUCATIONAL VISITS

The 1988 Education Reform Act states that no school journey or visit taking place wholly or mainly in school time, nor any excursion undertaken with children as a part of curricular work, has to be paid for by parents. This will probably mean that there will be fewer of these projects if they have to be financed by the school in whole or part.

However, a third party, a travel agency for example, may still organise such trips on behalf of the school as long as the members of the staff do not handle any of the financial arrangements. It would be as well to check your LEA policy on this, and also make sure that its insurance cover will still apply to trips arranged by outside bodies. Parents may also be asked by the school to make a voluntary contribution for their child to cover the cost of any outing, although no child may be barred from going on such a trip because his parents have not contributed. If too many parents refuse to make a contribution then the school has the choice of paying for the trip from school funds, or cancelling it due to lack of support.

If the school does embark upon an expedition, the chances of things going wrong increase sharply outside the environs of the school. To combat this the headteacher must draw up a written statement of intentions, to be observed on visits, whether for a day or longer. Such visits should be educational and at least one member of the staff must have visited the area in advance and reported back to the headteacher and any colleagues involved, before the planning stage.

There must be an adequate number of teachers, supplemented perhaps by parents and ancillary helpers. Some LEAs stipulate the ratio of adults to children on a school journey. In the case of juniors one to eight, or one to ten will be common. A visit needing a specialist helper, for example fell-walking or something similar, must have a suitably qualified person in the party.

The permission of parents must be obtained for any trip. Meetings should be held with these parents in advance. Any necessary insurance policies should be taken out.

On the visit itself, all adults and children concerned should know the exact purpose of the expedition and the contingency plans for any emergency, in case children should become detached from the main party. Maps, sufficient food and drink, and first-aid supplies should also be taken.

Record keeping

The headteacher, even in a relatively small school, cannot hope to keep all the details of the children in her head. Adequate records of the work and progress of children in our schools must be maintained. In 1989, the DES published *Records of Achievement*, dealing with investigations into secondary-school record keeping, and doubtless there will be guidelines for records in primary schools as well, just as most LEAs already issue their own cards for primary schools to maintain. Even so, it is incumbent upon all primary schools to keep their own, more comprehensive records.

Class teachers will keep their own notes of work accomplished, test results, and so on. The headteacher will need two main sources of reference for each child in the school, a record card and a record folder.

Record cards can be designed by the head. A local printer will run off

Side One

ST THOMAS' PRIMARY SCHOOL   Child's Entry Card

Name ....Date of Birth ....Date admitted:...............

Address

　　..

　　..

Tel: (Home)................................. Tel: (Other)....................................

a thousand at a time at very reasonable rates. A card about 14cm long and 9cm wide, used on both sides, will provide room for most records needed. These should then be filed in a tray or drawer, one class at a time. I have found the outline shown on page 51 to be satisfactory.

The other side of the card is used for reading and numeracy test results conducted each year, and any county or national test scores. There is space for the children's social, aesthetic and physical qualities to be entered.

The other record-keeping for the headteacher should be a folder in which examples of the child's work may be kept, so that the head may refer to it when necessary. Each teacher in turn, as the child passes through her hands, should be asked to include in a separate envelope inside the folder the following examples of the child's work in its last term in the class:

- 1 piece of creative writing
- 1 piece of handwriting
- 1 example of mathematics work
- 1 example of project work
- Any particularly outstanding piece of work which does not come under the above headings.

No one will expect the headteacher to know the details of the progress of every child in the school, but if the head can refer to the record card and the folder when parents ask for advice, these should provide enough material for the assessment asked for, when used in conjunction with the class teacher's opinion.

Organisation

The way in which the children in the school are organised into classes will depend upon the numbers, buildings and educational philosophy of the teachers and the head, but any arrangement should obviously be a sensible one likely to gain the approval of parents and governors.

The most logical arrangement is to place the children in separate classes, one for each different year of the age-range, from entry at the age of four or five years, to when the children leave at 11 or 12.

In a larger school there might be two or even three parallel classes for each age-group. This presents few problems. Where trouble does occur is when the children in the school do not divide up neatly into roughly the same numbers for each age-group.

There could, for example, be 43 second-year juniors, but only 18

THE CHILDREN

third-year juniors. This is where the headteacher, after consultation with the staff and governors, has to decide whether to introduce vertical grouping, in order to make all the classes the same size.

Vertical grouping means putting children of one age into a class of children of another age to make the class sizes more equitable.

For example, there could be three infant teachers in a school, but the infants could be spread in this way over the age-groups:

1st year: 36 children
2nd year: 20 children
3rd year: 25 children

In a case like this, the head and the teachers would have to decide whether to leave the children to be taught in their age-groups, but most schools would almost certainly decide that 36 first-year infants would be far too many for one teacher. One answer here would be to leave the third-year teacher with her 25 children, but to form two vertically-grouped classes from the first- and second-year children:

1st year children: 36 divided by 2 = 18
2nd year children: 20 divided by 2 = 10
= two vertically-grouped classes, each with 18 first-year children and ten second-year children
= 28 children in each class

Many teachers and parents do not like vertically-grouped classes, as this means that the teacher virtually has to combine two classes in one, and the younger children tend to suffer under such conditions. However, often numbers dictate that such a system has to be used. An HMI survey in 1988, showed that seven out of ten primary schools were using some form of vertical grouping of children.

Another method, but this time adopted by choice in some primary schools, is that of family grouping, in which a child might stay in the same class for the whole three-year period of its infant career. This arrangement is explained by Lorna Ridgway and Irene Lawton in *Family Grouping in the Primary School* (Ward Lock Educational, 1965):

> In a school where family grouping is used, a child remains in the same class, with the same teacher, for the whole period of his Infants' school life.
> Entering with a few (perhaps 5 to 8 other newcomers), he joins a class which will already contain 10 to 12 children of rising 6-year-olds, and a

further 10 to 12 rising 7-year-olds, all of whom have spent their Infants' school life up to that point in the same classroom taught by the same teacher (unless there have been staff changes).

Some heads believe that this approach gives stability to a child. It would be unwise to adopt such a method of grouping unless all the teachers involved believed whole-heartedly in its efficacy.

In some schools, the organisation of the children will be dictated by the design of the building. In an open-plan arrangement, for example, there might be less emphasis on formal class work and more on group activities.

If the school is large enough to have more than one class in each year, the headteacher is faced with the problem of how to group the children within the year. Are the children going to be streamed, with the most able in one class, the more average in another and a group of slow-learners in the third, for example? On the whole, practice today and parents do not favour this form of grouping. The high-achievers may benefit from such an approach, but the average and below average pupils seem to benefit more from being in mixed-ability classes. Provision must be made, of course, for the slower-learning children to receive extra help, or to be withdrawn for certain periods, perhaps by the headteacher, for special teaching.

If the design of the building allows it, children may always be brought together for various forms of team teaching. Several classes may be 'set' once a day for, say maths, with one teacher taking the more gifted and another the less able. This enables specialist teachers in the primary school. If there are more than two classes in a year group, the whole year could be brought together once a week for project work, with the different teachers taking responsibility for various aspects of the topic, perhaps with the aid of the headteacher as well.

Provision must be made for the gifted children in our schools, as well as the slow-learners. This will involve much care and sympathy on the part of the teachers, and a great deal of individual attention. Again, the head could help out by taking the high-achievers for informal sessions once or twice a week, in an effort to broaden their horizons and encourage them to look beyond the immediate curriculum.

In his book *Teaching Able Children* (Kogan Page, 1988), Tom Marjoram stresses that the gifted need scope in order to relate abstract ideas, innovate, pursue notions and topics in depth:

> In planning curricular provision for the gifted, then, we must retain breadth. Every gift can only be fully illuminated against the backcloth of the whole

of human knowledge. Nevertheless, within that broad, coherent core, there needs to be much choice and variety.

REGISTERS OF CHILDREN
The school secretary should keep an entrants' book in which details of each child joining the school and the names and addresses of parents should be entered, and another book with the names of those children waiting to join the school when they are old enough. The class teachers will keep records and registers of their children, and the head should always have several typed, up-to-date class lists close at hand.

One of the most effective ways of keeping an eye on the distribution of children throughout the school, and those waiting to enter, is to have a series of long wooden or metal card-holders hanging vertically from the office wall. There should be about 35 slots in each holder, so that cardboard strips bearing the names of individual children may be inserted. There should be one long holder for each class, and several additional holders for the year-groups waiting to be enrolled. If a different colour of cardboard is used for each year, the names of the children in each class may be inserted and used for quick reference.

Preparing children for secondary education

Most children make the transition from primary to secondary school with little difficulty. Indeed, the 1988 ILEA survey, *Secondary Transfer Project*, came to the conclusion that the number of children enjoying school doubled as they moved from the top year of the junior school to the first year of the secondary school.

As one of the last services that we can perform for our children, we should ensure that links are built with the secondary schools which will be receiving our leavers, and that the children get the opportunity to visit these schools and meet the teachers. There are several ways to link with secondary schools.

- Arrange for secondary-school teachers to come in and talk to the leavers' classes.
- Try to arrange with the secondary-school heads a series of open days for the primary-school leavers at these secondary schools.
- Try to organise one or two joint projects with the secondary schools, involving the primary-school leavers.
- Try to bring in secondary-school pupils to meet the leavers and work with them occasionally.
- Exchange staff with the secondary schools for brief periods.

Child abuse

School staff members have a duty to be on the lookout for any signs of child abuse. These signs may reveal themselves in the form of cuts, burns and bruises when the children change for games. Sexual abuse may exhibit physical signs, or may be indicated in a change of personality on the part of the child – lack of confidence, desire for attention, etc. Teachers should also be on the lookout for signs of neglect in the clothing of the children, abnormal signs of hunger, and so on. If a teacher suspects any of these things she should report them at once to the head.

If it is possible to do so with tact and sympathy without distressing the child, the headteacher could question the child in order to find out the reason for the change in the pupil's appearance or behaviour. It is *not* the duty of the headteacher to pursue this matter with the parent or guardian concerned.

If the headteacher believes that there is a suspicion of child abuse, or if the child himself confirms such a thought, she should at once follow the guidelines issued by every LEA. These will include the name of a responsible member of the Education Department, who must be contacted at once.

This officer may deal with the matter himself, or may instruct the head to contact the Social Services Department, or the Police, or a professional care organisation.

Unless the official of the Education Department suggests it, which is unlikely, parents do not have to be informed of any referral of this nature. The care organisation concerned will take the matter up. The headteacher is advised, however, to contact her chairman of governors and inform him of the situation. A comprehensive note of the condition of the child, any complaint made, and the action taken by the school, should be made, and one copy sent to the named official of the Education Department, while another copy should be filed.

Police and the school

Many local police authorities have school liaison officers who maintain good and friendly relations with the schools in their areas. A number of LEAs have their own policies about the admission of uniformed police officers on to school premises, and these of course must be observed.

The prudent headteacher should make every effort to maintain good relationships with the local police representatives, and many primary schools have police officers running clubs, coaching football teams, giving talks on road safety, and so on.

The school has a duty to assist the police in any investigation of a crime. Names and addresses of pupils may be disclosed to the police, if they are requested. If a police officer should visit the school with a request to interview a pupil about some suspected offence, the headteacher should request the officer concerned to wait until the parent or guardian responsible for the child has been sent for and can be present at the interview. A child may only be interviewed on the school premises with the permission of the head and in her presence, and the child must agree to be questioned. It is far better to await the arrival of a parent or guardian in almost every case.

Supervision after school

Care should be taken not to release children from the classrooms before the official ending of the school day. A child who is injured or hurt on the road because he was released too soon would have a claim against the school through his parents.

Teachers may be regarded as being responsible for children for a period of some 15 minutes after the end of afternoon school, although of course if parents or taxis or other forms of transport have not arrived to pick up children the headteacher is still responsible for the safety of the children until they are taken away from the school premises. The headteacher is not responsible for the supervision of children using transport once they have left the school, although she has the right to investigate and if necessary punish bad behaviour on that transport.

DETENTION

The use of detention after school as a punishment for children may be justified if the teacher concerned can show that she considered that form of punishment necessary. Detention for primary-school children is, however, an emotive issue. It should certainly not be used if it means a child having to walk home alone, or after traffic wardens have left their posts, or if a long distance has to be travelled. Parents should certainly be informed if their children are going to be kept in after school, and told what time to expect their children. On the whole detention is a punishment which raises more problems than it solves in primary schools.

Summary

The children are the most important part of our job. It should be our main task to establish a good relationship with them by letting them see that we are hard-working, efficient and fair organisers of their school lives, work and surroundings. We should be familiar to them as teachers and helpers of their teachers.

Our shop-window should be the school assembly. This should enable us to make contact with the children and the staff on a daily basis.

There should be a brief but comprehensive code of discipline for the school, known to everyone and containing more rewards than punishments. The children should be well cared-for and properly supervised as long as they are in our charge. The total administration of the school should be geared to the needs of the children and their teachers.

These are some of the ways in which we can serve the children efficiently:

- Know the children. Make it a matter of priority to know every child in the school by name and something about each child.
- Be known by the children. Get around the school frequently.
- Gain the respect of the children by working for it as a teacher, administrator and person.
- Be fair and consistent in dealings with the children. Try to give reasons for what you do.
- Always have time for the children.
- Spend an hour a day teaching and another 45 minutes helping a teacher.
- Liaise with the local secondary schools.
- Have a firm code of discipline which includes dealing with bullying.
- Have a written code of supervision for all non-teaching sessions of the school day.
- Check the building and plant regularly.
- Have a comprehensive fire-safety policy and hold regular fire drills.
- Maintain well-stocked first-aid boxes throughout the school.
- Keep adequate and regularly updated records on all children.

Self-assessment

The ethos of the school will stem from the relationship of the headteacher with the children. We should ensure that we place the education and welfare of the pupils above everything. Are they happy and well cared for? Are they well-organised? Are they properly motivated?

QUESTIONNAIRE
The headteacher's image
What sort of impression do you think you make on the children of your school? Do they regard you as aloof, or an integral part of the life of the school? If you feel that you can improve your image with the children, draw up a plan to help you achieve this aim.

The headteacher as teacher
Do you feel that you teach too much, too little, or just about the right amount of time? Is your assembly the show-place where you put your ideas and aspirations into practice? Do you feel that any changes might be indicated here?

Rewards and punishments
Examine your code of discipline. Do rewards outweigh the punishments? Are all children aware of any basic rules? Does the school have an exclusion policy?

Care and supervision
Does your school have a detailed code of supervision covering the entire school day? Are there any changes you feel should be made to it?

Record keeping
How effective are your records of children's work and progress? Do they include sporting and aesthetic interests and abilities as well as educational achievement? Could you put your hand on an accurate and up-to-date assessment of the work of every child in the school?

Organisation
Does your system of organisation fit the needs of your school, the abilities of the staff and the requirements of the children within physical constraints of the building? Is provision made for both the slow-learners and the gifted among the children? Are there any changes you might contemplate?

---

continued overleaf

Child abuse
Do you know whom to contact in the Education Department and what other steps to take if you suspect that a child in your school is the victim of abuse?

Chapter 4
The Staff

Motivating and orchestrating the varying degrees of teaching talent and application we have at our disposal is difficult, but has long been regarded as one of the main functions of the headteacher. Thirty years ago, Her Majesty's Inspectors were pointing this out in their publication *Primary Education* (HMSO, 1959):

> To encourage and guide the best that each can give and cultivate a sense of unity among all who work in the school, from the young and untried to the older and more experienced, and from the less competent to the distinguished, calls for the utmost patience, good sense, humour, humility and sense of purpose.

In order to do this, the modern primary school headteacher has to divide her responsibilities for the teaching staff into a number of overlapping sections – recruitment, appointment, deployment, treatment, training and assessment.

Recruitment of teachers

Until the arrival of short-term contracts, every headteacher must ensure that she is familiar with any specific regulations concerning the appointment of staff laid down by the LEA, or contained in the regulations of the school. Governors should be involved in the appointment of staff from the very beginning of the process. The headteacher should see to it that the governors are kept sufficiently in touch with the life of the school to be aware of its strengths, weaknesses and needs as far as personnel are concerned. The headteacher and teacher governors should have a good idea of the educational qualities needed on the staff, but the other members of the governing body should possess between them the width of experience and the judgement of character to be able to seek out the personal qualities needed in the successful applicant.

DRAFTING THE ADVERTISEMENT

Once the initial recruitment has been decided upon, most governing bodies will leave it to the headteacher to draft the job advertisement and the description of the post to be filled. Whenever possible, as a part of her in-service training, the deputy should be involved in this.

The advertisement should be as short and succinct as is compatible with clarity. There is a great temptation to overload the advertisement in an effort to attract the right candidates. This is not the stage for such fine tuning. We need to get as many replies as possible to our overtures; there will be time to cull later. If we are over-meticulous in our stated requirements we risk frightening off perfectly reasonable candidates. Even if we wish to appoint someone to assume responsibility for computers or football, or something of that nature, in primary schools there is still a case for compiling an all-purpose type of advertisement: *Teacher for Juniors required. Please state interests.*

In this way we should still get the football and computer enthusiasts replying, but it is also possible that an outstanding candidate without these particular strengths may also show an interest in the post. In such a case the governors might decide to short-list the non-specialist along with the others.

One piece of research came up with the finding that most advertisements for teaching posts consisted of idealised self-portraits of those drawing up the lists of requirements, which might explain the plethora of demands for candidates possessing qualities of personality, leadership and scholarship which could only be met by the joint appointment of Mother Theresa, Mike Brearley and Bertrand Russell. Nor should we describe our schools in glowing terms in the advertisements. It will be far better to allow the candidate to judge for herself.

JOB DESCRIPTIONS

Once the advertisement has been drawn up and placed nationally and locally as devised by the LEA, the headteacher should draw up a detailed job description to be sent to applicants along with the official application forms.

A job description should occupy about an A4 sheet of paper. The description should contain a few basic facts about the school, the post and the area:

- Brief description of town or area

THE STAFF

- Types of accommodation available with price-ranges
- General description of school and environment
- Title of post and salary scale
- Number of children on roll
- Class to be taught and number of children in class
- Person to whom teacher will be responsible
- General grouping arrangements within school
- Skills and qualifications looked for
- Other responsibilities
- Any other relevant information
- Closing date for applications.

When the completed application forms start to come in, it is usually the task of the headteacher to make out the first long-list, although members of the governors' sub-committee may take a hand.

Then the governors will select a short-list of suitable candidates and write to their referees, seeking additional information. When this has arrived the governors may decide whether they still wish to interview the candidates originally selected, or go back to the long-list again.

References are essential to the process of choosing candidates for the short-list, but must be treated with caution. Unfortunately it is a fact that some members of our profession will praise a teacher to the skies in order to be rid of an incompetent colleague. There is little that can be done about this. One has to rely upon the honesty of the referee. Referees should be asked to answer certain specific questions. It is always possible that the referee has forgotten to mention a particular quality, but its omission could also mean that the referee prefers not to dwell upon that aspect of the candidate's performance. In either event, it is best to make sure.

The ability of an applicant to get on with colleagues and children, enthusiasm and dedication are all attributes which may better be judged from the considered report of a colleague than discovered at an interview. Health and attendance should also be queried. These may not figure very highly in academic treatises on staff selection, but in the real world a teacher who is likely to have a lot of time off is going to be a liability in a busy primary school.

The following is a list of points to be looked for in application forms and accompanying letter:

- Are the forms filled in correctly?
- Is the accompanying letter literate and sensible?
- Is the candidate suitably experienced?

- Are the candidate's qualifications adequate for the post?
- Does the candidate possess any additional qualifications – coaching certificates, etc?
- Has the candidate attended a reasonable number of relevant courses?
- Has the candidate any work experience outside teaching which might be of use?
- Does the candidate's age suggest that she will fit in to the staff pattern?
- Does her record of attendance seem reasonable?
- Are her references good? Are most of them from heads or teachers who know her work?

INTERVIEWING CANDIDATES

Whenever possible, candidates on the short-list should be invited to look round the school before they are due to meet the governors. It may help if the teacher-representative on the governing body is deputed to show the candidates around the school and answer their questions on a practical level, as one teacher to another. Some heads, especially if the advertised post is for a graded position, ask the candidates to spend the morning in the school, sometimes even teaching groups of children, but this seems to be asking much of colleagues who must already be under a considerable strain.

A number of LEAs like to see the applicants on the morning of the interview, in order to weed them out still further, before sending the survivors on to the school in the afternoon to face the governors. I once attended such a gathering myself. In the course of this rather odd morning, individual candidates, as they were found wanting, were asked to take no further part in the proceedings, until only four were left to go on to the interview. Such situations are artificial, stressful and, in my opinion, not very effective. The orthodox interview has its limitations, but until something better is devised, most schools would do well to adhere to this method.

It is a good idea to try to restrict the interviewing panel to a maximum of six. Too many interviewers are not only off-putting but they tend to get in one another's way and muddy the waters. However, most governors like to become involved in this important stage of the selection process, so the panel may be a large one.

At the outset the chairman should go through the procedure to be followed. This may be done while the applicants are being shown round the school. Care should be taken in advance to avoid the duplication of questions. Governors should be encouraged to make their questions as open-ended as possible, so that candidates are not forced to make terse replies. It is usually a good idea for one of the interviewing panel to ask the applicants what they think of the school. Their answers should at least reveal how observant, perceptive and tactful they are.

Interviews should be conducted in as pleasant and non-threatening a manner as possible. It is essential that the interviewing procedures are manifestly fair and impartial. There must be no discrimination against candidates. This is an area in which not only must justice be done, but must manifestly be seen to be done.

DISCRIMINATION LEGISLATION
Sex Discrimination Act 1975
A person discriminates against a woman in any circumstances relevant for the purposes of this act if:
(a) on the grounds of her sex he treats her less favourably than he treats or would treat a man,
or
(b) he applies to her a requirement or condition which applies or would apply to a man but:
   (i) which is such that the proportion of women who can comply with it is considerably smaller than the proportion of men who can comply with it, and
   (ii) which he cannot show to be justifiable irrespective of the sex of the person to whom it is applied, and
   (iii) which is to her detriment because she cannot comply with it.
These provisions are to be read as applying equally to the treatment of men.

Race Relations Act 1976
A person discriminates against another in any circumstances relevant for the purposes of any provisions of this act, if:
(a) on racial grounds he treats that other less favourably than he treats or would treat other persons; or
(b) he applies to that other a requirement or condition which he applies or would apply equally to persons not of the same racial group as that other but:
   (i) which is such that the proportion of persons of the same racial group as that other who can comply with it is considerably smaller than the proportion of persons not of that racial group who can comply with it; and

(ii) which he cannot show to be justifiable irrespective of the colour, race, nationality or ethnic or national origins of the person to whom it is applied; and
(iii) which is to the detriment of that other because he cannot comply with it.

THE INTERVIEW

At the interview, each candidate should be allowed twenty or thirty minutes in front of the committee. At the end of this time she should be asked if there are any questions she wants to ask. The interviewers should bear in mind the following points:

- Does the candidate match up to the impression created in the application form and references?
- Does she appear to be well-motivated and a 'self-starter'?
- Does she seem well-organised and confident without being overbearing and bumptious?
- Does she appear fit and enthusiastic?
- Are her personal qualities such as to make her seem a suitable candidate?
- Does she appear to have interests beyond the school and classroom which will make her a more fulfilled and versatile person?
- Does she appear to have qualities of initiative, judgement and originality?
- Does she seem genuinely interested in children?
- Does she seem the sort of teacher the parents would respect and get on with?

When all the applicants have been seen, the chairman should sum up the merits of each candidate placed against the needs of the school before throwing open the meeting to discussion. Allowance should be made for the natural diffidence of some candidates.

The part played by the headteacher in the selection procedure will depend upon her natural inclinations and the relationship she has with the governing body. Some heads are regarded as professional consultants by the governors; others take a full part in the debate, or even try to run it. On the whole, the more successful appear to be those of a Machiavellian turn of mind, who can get their way by stealth. The vast majority of governors would hesitate to appoint a teacher who did not have the approval of the head. If this should happen, however, try to accept the *fait accompli* gracefully. After all, the governors may be better judges of character than we are!

AFTER THE INTERVIEW
When a decision has been reached, the successful candidate should be called in and offered the post. While she is receiving the congratulations of the governors, the headteacher has the task of commiserating with the remaining candidates.

This calls for tact and discretion. It is certainly no consolation to the disappointed applicants if the head is patronising and over-ingratiating or uses the occasion to conduct a spontaneous seminar on interview techniques and the deficiencies of the individual candidates. The best that a sympathetic head can do is thank the teachers for coming, wish them well in future applications and make sure that they have all been issued with expense-claim forms.

Deployment of teachers

The deployment of teaching staff is not usually a complicated matter in the average primary school. In most cases each teacher will be placed in charge of a class. The main problem facing the head lies in providing her colleagues with opportunities to work together in teams, exchange classes in order to work regularly with children of different age-groups, and practise any specialist skills throughout the school.

EXTENDING THE EXPERIENCE OF TEACHERS
It will prove helpful if two or three teachers can be given opportunities on a regular basis to combine their classes for such activities as music, games, maths sets, etc., all working together in the hall or playground or on the sports field.

This will enable teachers to organise larger groups than they normally work with and gives them the chance to observe the techniques of colleagues.

Similarly, the headteacher, in consultation with the staff, should arrange for teachers to exchange classes two or three times a week, for one teaching period at a time. Again, it will be a useful exercise in administration for the teachers to arrange what they are going to do with one another's classes. By moving up the school, the infant teachers will see how their former charges are progressing, while the teachers of the older children will then encounter the boys and girls they will be looking after regularly in a few years' time.

An effective way of helping teachers to work with the children they do not normally teach, and at the same time to widen the range of the school's activities, is to timetable the whole school for a regular weekly session of clubs and what might broadly be termed cultural activities.

This could be for one 45-minute period a week, perhaps for the last period after play one afternoon. Parents may also be brought in to help.

The headteacher could take all the infant children for a combined music period or a television period with the necessary follow-up activities. This would enable the infant teachers to join the activities in the junior school.

Every term, the composition of each group should be changed, so that over the period of the school year, every child could join three different activities, each time with a different teacher. Activities could include computers, choir, Brownies, drama, country dancing, art and craft, gymnastics, needlework, Cubs, cookery etc.

Responsibilities

It is essential that teachers be made to feel that they have a personal stake in the school. This can best be done by involving them at all levels of responsibility, giving them the opportunities to make decisions, with the headteacher delegating as much as possible. Within the directed-time hours there is plenty of scope for teachers to take part in many aspects of responsibility. For example,

- Overall responsibility for a subject specialism throughout the school, for example, language, maths, music, design and technology etc
- Sports and games
- Display
- Computers
- Assemblies
- Library
- Concerts and drama
- Health and safety

All except the smallest primary schools will have incentive allowances at Grade A and Grade B to allocate. These should be given for tasks involving high levels of responsibility and decisions, for example,

- Responsibility for infants
- Ordering and maintaining stock
- Liaison with secondary sector
- Responsibility for juniors
- Home–school relationships
- Special needs

Most teachers will prefer to decide for themselves how they will tackle their special responsibilities. With weaker, less confident or inexperienced colleagues, it is usually advisable to go through their proposed tasks with them and suggest sets of definitive objectives – immediate, mid-term and long-term. In the case of the less able teacher, the objectives suggested should be within her capabilities.

A teacher given the task of the revision of reading schemes could have as her objectives:

- Immediate (this term):   Discuss with colleagues any alterations needed in present schemes.
- Ascertain amount of money available to purchase new books. Catalogue present supply of reading books.
- Mid-term (next term):   Colour code existing books and any new ones.
  Set up system for parents to come in and listen to children reading.
  Arrange physical areas of the school to be available for parents to listen to children.
  Order supplies of new books as money becomes available.
- Long-term (by end of school year):   Develop home–school reading scheme.
  Arrange system of recording and testing children's reading.
  In collaboration with colleagues, devise and write down the school's reading policy.

Once you have delegated tasks to teachers, let them get on with the work. Do not hover at their elbows or try to second-guess their decisions, unless matters are going badly wrong or inertia seems to have set in. Expect each teacher to report to you regularly about her area of responsibility otherwise stay away from the detailed work. There is nothing more annoying than a head who will not let go of the reins, except one who keeps gathering them in again!

Treatment of teachers

The organisation, training and appraisal of our colleagues can be plotted and planned. Living with them for up to eight hours a day demands a more pragmatic approach and a certain amount of flexibility. As we all know, the staff of any primary school is not made up of colourless automatons, but of living people, with different backgrounds, personalities, opinions and aspirations. Every school has its store of anecdotes of the wide diversity of teachers' characters and foibles, but less conspicuous and much more important are the countless acts of love and skill performed in thousands of classrooms all over the country as caring teachers, with many minute, detailed touches, put together the fabric of education.

It is part of our responsibility as headteachers to notice these contributions and to let our colleagues know that we value them. It is our duty to support and protect our colleagues, and at the same time to help motivate the less gifted ones, so that we nurture and cherish what is good in our schools and seek to improve what is less good. Our approach to all teachers should be geared to their own individual personalities, but should at all times be gentle and supportive, but with quiet insistence on certain standards being maintained. No matter how sorely tried or misunderstood we may be, we should try to keep our humour and our dignity.

COMMUNICATIONS

Perhaps the two most important ways in which we can make useful and productive contact with our colleagues on the staff are by understanding them and by making sure that we are able to communicate with them.

We will gain the confidence of our colleagues and get to know them more quickly if we are respected by them as adequate leaders, organisers and teachers. We should know what is going on in the school and be aware of educational developments nationally and locally and be able to translate these events into practical aims for our schools. It is important that we know what sort of assistance is available in our own areas, so that we can bring in inspectors and specialist teachers and experts to help the school in certain situations.

We must also seek to earn a reputation as facilitators who know how the local and national administrative services work. It is far better to be thought of in the 'Jim'll fix it' context than to hear the ultimate indictment from colleagues of 'It's no good going to her!'

In all but the largest primary schools we should try to talk to each teacher informally every day. This will only take a few minutes in each classroom before or after school. It goes without saying that we should establish a reputation within the school of utter discretion and reliability. Under no circumstances should we discuss colleagues behind their backs, unless it is the head's policy to share all problems with the deputy.

We should know our colleagues well enough to guess when it is time to leave them alone, and when to act as a buffer between them and the outside world. If a teacher is going through a bad patch personally we should make it our business to know this, then we must decide whether to offer assistance or let her work out her own problems, as long as the school and the children do not suffer as a result.

We should be friendly and approachable but should respect the privacy of our colleagues and remember that they can never really relax in our presence. While we should use the staff room in an informal manner, so that we may be seen and approached, we should also take care to vacate it for long periods at playtime and lunchtime each day, in order to let the teachers relax, even if it is only to complain about our latest errors of judgement!

If as headteachers we expect our colleagues to be frank with us, then we should ensure in turn that communication within the school is a two-way process. Clearly there will always be some confidential matters which we should keep to ourselves or impart only to the teacher concerned. All the same, as far as possible we must endeavour to operate an open-door policy of trust within the school, involving our colleagues as much as possible by allowing them to share in the concerns of the school as a whole.

There are a number of ways of disseminating information. Some of them are more effective than others. The most obvious, word of mouth, has its pitfalls. I was upbraided by a distressed colleague because I had fallen into the habit of raising points of general interest and discussing them informally in the staff room at lunchtime, when this particular teacher had gone home. I had assumed, correctly that the other teachers would pass on any relevant matters to her, but as the teacher pointed out, I was treating her differently and she felt left out.

The staff notice board also has its uses as a focal point for information, but here again there are drawbacks. In most schools the staff room is used for a variety of purposes by children, parents and visitors. This means that confidential information cannot be displayed where other eyes may see it.

Staff meetings are generally regarded as the ideal occasion for an exchange of ideas and views. These tend, however, to be formal occasions, held perhaps twice a term, when a formal agenda of important curricular and organisational matters may be examined in detail, notes taken, decisions made and plans laid for the future. Some heads attempt to solve this problem by holding staff meetings more often, sometimes every week. Even the most dedicated and idealistic teachers draw the line at this. They are prepared to meet in small groups regularly to sort out pressing problems, but dislike too many formal staff meetings.

The simplest method of keeping teachers in touch with what is going on outside their classrooms is for the head to take the time and trouble to prepare and circulate a confidential staff newsletter every two or three weeks, as long as it is remembered that this letter is supposed to complement our personal contact with everyone, not replace it. The letter could look like this:

Staff Newsletter Confidential July 10th

Please tick and pass on; Class 1 2 3 4 5 6 7 Head

Visitors: Mr Jones the RE Inspector proposes visiting the school on Monday 14th July at 9.30 am
Mr Smith, a parent governor, would like to visit classes on Friday, 18th July from 9 am until 10.30 am

Children: Stacey Brown in Class 6 will be leaving on Friday. The family is moving to London.
Mrs Green, mother of Anne in Class 1 and Brenda in Class 4, and her husband have parted. Mrs Green has asked us to look out for any signs of distress on the part of the children.

Staff: Mr Black will be away on Tuesday 11th July on a science course. A supply teacher has been booked for Class 5.

RELATIONSHIPS

We must work constantly at our relationships with our colleagues. This means that we must be extremely careful what we say, in case the wrong construction may be placed on quite an inconsequential remark. We should also take care to treat all teachers in the same way. For a head to have a 'favourite' is quite as damaging as a class teacher favouring one particular child, no matter how gifted.

We have to take care not to operate behind the backs of any of our colleagues. One of the worst tangles I ever got myself into was when one teacher complained to me about the behaviour of a colleague, but begged me not to reveal that she had made the allegation. Foolishly I investigated the complaint without involving the original teacher. It later transpired that the two colleagues had been at loggerheads for years. By the time I had worked this out I had offended just about everyone on the staff and brought all sorts of trouble down upon my head when both teachers, combining for the first time in their lives, protested about my actions to the LEA and their union. What I should have done, of course, was to have insisted straight away on the first teacher making the complaint in the presence of the second while the three of us tried to sort the matter out. If the original complainant had refused to do so, then I should have told her that the matter was closed.

We should also make it a practice always to be around. There are many temptations to absent ourselves from the school on various errands, but a head who is constantly away on courses or at meetings cannot command the respect of the staff.

Appraisal of teachers

The prospect of being judged and perhaps found wanting by others has probably thrown the staffs of primary schools into more of a flutter of apprehension than any other clause of the 1988 Education Reform Act. In fact of course, we have all been appraised regularly from the beginning of our careers. Children, parents, colleagues, even members of the local community, all have their own ideas as to whom the 'good' teachers are in a school, and whether or not the head is doing an adequate job.

This is all the more reason to have an efficient, fair, supportive and non-threatening system of staff appraisal. In this way many myths may be dispelled and teachers will benefit from examining their own methods in some detail. This, of course, is the crux of the matter. The best and most effective form of appraisal is that self-assessment carried out by the teachers themselves, aided by a sympathetic headteacher who knows the staff well and has the trust and confidence of the teachers.

When the appraisal process is launched, the head should give all members of the teaching staff plenty of advance notice, provide details of the logistics of the operation, and outline the areas of professional competence which the teacher will be asked to analyse and then discuss with the head.

This is a list of likely steps in an appraisal:

1. First meeting between head and teacher to arrange general time-sequence.
2. Teacher is given time to think about the appraisal, check on job description, school curriculum, timetable, etc.
3. Head and teacher meet again to clear up any practical matters and go through the areas to be appraised. These areas will be studied by head and then discussed with the teacher.
4. Headteacher spends time watching teacher in the classroom.
5. Headteacher and teacher meet. Teacher asked to appraise herself. The head then says what she thinks about the same areas. Head and teacher agree on any weak points and agree to meet again after they have both thought about how these areas might be improved.

6. Head and teacher meet again to set targets for teacher in any areas where these are considered necessary.
7. Written appraisal statement issued to teacher.
8. Head arranges in-service training, visits to other schools, etc., designed to help the teacher in areas of weakness.

The headteacher should write down the areas of professional competence which will be examined, and give the list to the teacher, so that she can prepare a self-assessment form to complete during the appraisal process. The main areas of professional competence to be assessed will be:

1. Approach: variety of teaching methods and materials employed with enjoyment, enthusiasm and energy.
2. Planning: lessons prepared in accordance with syllabus, taking into account abilities of class.
3. Teaching: lessons structured, interesting and up-to-date.
4. Follow-up: children's work marked, assessed, tested and recorded.
5. Attitude: establishment of good relationships with children, involving courtesy, humour and respect for learning.
6. Control: good order maintained among children, but teacher also able to relax among them.
7. Attendance: good attendance and punctuality.
8. Enthusiasm: a full part taken in out-of-school activities.
9. Sensitivity: an understanding of children and ability to help them in pastoral side of school work.
10. Contact: gets on well with colleagues, parents, etc.

When the teacher has been observed over a reasonable period of time, has carried out a self-appraisal exercise and has had time to think about the whole process, the head should carry out the next stage in the assessment and go through the self-appraisal form with the teacher, finding out if they agree on the main points in it. The head must make it plain that the appraisal is intended purely to help the teacher, not to threaten her, and that it is the performance, not the person which is being discussed.

What is good in the teacher's style should be praised, and that which might be improved should be discussed frankly but gently. Most of us know our own strengths and weaknesses and will be prepared to do something about them as long as we do not feel threatened and if we feel that the improvements being asked of us are capable of being obtained.

When the areas of weakness, if there are any, have been agreed upon and debated, the headteacher and the teacher together should start to draw up a set of goals to be aimed at. This should take the form of a few structured, practical steps for the teacher to take over a stipulated period of time. Agreed targets on lesson preparation might be:

1. Study school curriculum for English, maths, science.
2. Work out set of detailed objectives in these subjects for children to achieve by the end of each month in the Spring term.
3. Set these objectives down in writing.
4. Teach the children using these notes you have made.
5. Write a detailed analysis of how close you were able to keep to these objectives.
6. Discuss results with headteacher.
7. Follow the same procedure for Summer term, but this time add history, geography, design and technology and music to your list.

If the matter of appraisal is approached sensibly and with mutual tolerance, it can be an interesting and productive experience. It will certainly help to clear the air. Sometimes for the first time, teachers might know what is expected of them and whether or not they are living up to expectations. At the same time, if only obliquely, the headteacher will also be appraised every time she conducts an interview with a colleague, because we shall need to know if we are backing up our staff properly and providing them with adequate resources and encouragement.

It is generally agreed that the most successful forms of appraisal are those which arise naturally from the day-to-day professional life of the school and the on-going encouragement of the teachers by the head. As a rule, it is only when an appraisal process is introduced in an insensitive and inconsiderate manner that the teachers fail to respond to the opportunities provided for self-development.

APPRAISAL OF HEADTEACHERS
It is the responsibility of each LEA to devise a system of appraisal of headteachers within the agreed national framework. It is likely that heads will be appointed by their peers, appointed or seconded for the purpose, and that the views of governors, parents and the LEA itself will be taken into account.

Training of teachers

Our school's log-books go back over a hundred years. Once, when I was in a particularly bumptious mood and trying to initiate too much at once, a colleague who had been using the logs for a history project, gently pointed out that just about every new headteacher since the opening of the school in the 1870s, had started his or her term of office by making an entry in the log to the effect of 'Today I took over command of the school. I found it in a parlous state and resolved to train up the staff to my standards. ...'

There is a danger that in order to justify our existence or to impress others we will try to alter too much too soon. If things are going well we should make any necessary changes in the teaching methods of colleagues slowly and almost imperceptibly.

WAYS OF TRAINING TEACHERS

There are four main divisions of teacher-training within a school. The first is the individual training of colleagues as a result of targets set after an appraisal exercise. The second is the training of teachers by one another, when subject specialists or colleagues with specific skills are timetabled to help other members of the staff in their areas of expertise. This is known as the cascade method, because the teachers who have learned new skills should in turn be able to pass on their newly-acquired abilities to others.

The third method of in-school teacher training is the daily on-the-job instruction carried out by the head. This should take place as the head helps out in the classrooms.

The fourth type of teacher training in schools is the specific programme devised by the head and her colleagues to cover the five so-called 'Baker' days which are to be devoted to in-service training each year. These programmes are often allied to training work developed by local education authorities.

The first method of training by setting specific individual training targets should begin by assuring the teacher that the goals being set are concerned with what the teacher does, not what she is. She is being judged professionally, not personally. Then after appraisal the headteacher should:

- Set specific targets and time-limits.
- Allow teachers to observe other teachers in the school and at other schools.
- Seek to motivate the teacher by discussing career prospects and how these may best be served.

THE STAFF

- Arrange for teacher to meet colleagues from school or from other schools who may have similar problems.
- Use the services of local inspectors and advisers.
- Consider the possibility of sending the teacher on a full-time course if an appropriate one is available.
- Consider a change of class within the school for the teacher.
- Take the class yourself with the teacher regularly.
- Support and counsel the teacher constantly.

A recent HMI report was critical of the fact that in most schools not enough use was made of post-holders and other specialists on the staff to train other teachers. It is a fact that in primary schools, if we are not careful, we tend to get bogged-down in our own classrooms. One method of avoiding this is to make sure that every few years a teacher gets a different age-range to teach. Another way of enlarging the horizons of the staff is to use all the teachers to pass on their skills and enthusiasms to their colleagues.

Headteachers helping specialist teachers to train colleagues should:

- Know the specific abilities and interests of colleagues and encourage them to up-date and develop them.
- Write down specific job descriptions, so that teachers know the areas for which they are responsible.
- Organise classes in groups of two or three occasionally, so that teachers can watch a specialist at work.
- Timetable yourself to take classes regularly to free colleagues to go from class to class training others.
- Arrange in-service courses in directed time or on Baker days, for specialist colleagues to train others.

As the headteacher joins each class in the school on a regular basis she will be able to observe colleagues in action and be in a position to make suggestions as to how individual teachers may be able to improve in certain areas. This should never be done in the presence of children, but in a one-to-one situation with the teacher after school.

The third method of training where the headteacher ascertains areas in which the teacher might need help in the classroom could cover:

- Layout: are the wall-displays attractive, relevant to the work being done and changed regularly? Is the work on display child-based rather than teacher-based?

- Organisation: do the children enter the classroom and go straight on with some work? Do they know the routine for obtaining pencils, paper, books, etc.? Are they aware of the main timetable divisions each day?
- Preparation: is the preparation always in hand, with materials for lessons ready in plenty of time?
- Lessons: are these structured, with an interesting beginning, revision of previous knowledge and then an introduction to fresh skills? Does the teacher check regularly that she is holding the interest of the children? Does she use her voice well and summarise points at the end of the lesson?
- Apparatus: does the teacher make good use of all available back-up materials and audio-visual aids, like radio, television, computers, etc.?
- Relationships: does the teacher have good relationships with the children, maintaining good order with a pleasant and friendly teaching style?
- Curriculum: is the teacher following the broad outline of the school curriculum and devoting sufficient time to the foundation and basic subjects?

BAKER DAYS

The greatest emphasis on the training of teachers today is the five Baker days provided for the purpose, together with additional in-service training provided by the local education authorities. This In-Service Education and Training (INSET) has been funded by the Grant-related In-Service Training and Development (GRIST) scheme and its successors.

On the whole, the five additional training days and any further time devoted to this INSET work is being used in schools for the professional development of the staff as a whole, and not for the training of individuals, although some heads may devote a proportion of the time and money available for this purpose.

The headteacher should ascertain how much money has been allocated for in-service training at her school, and then meet with the teachers to decide upon the training priorities. It may be decided that the whole school's staff needs updating on computer activities, or specific aspects of curriculum development, or health and safety practices. When this has been decided, the headteacher has to decide how to use the GRIST funding for the benefit of the school and which days should be devoted to this INSET work.

Three of the Baker days are usually allocated to the first days of each term. The teachers come in for the training on the day before the children start back. It is the placing of the other two days which causes concern. If they are placed at the end of a term the teachers are often too exhausted to benefit from them. On the other hand, if a head closes a school during term time for two days of training, parents often object to the disruption of their children's education and the personal inconvenience caused.

There are a number of possible uses of GRIST funding and INSET days:

- Pool resources with other schools in area for a day or days of training.
- Employ supply teachers to allow staff to visit other schools.
- Hire outside experts to come in and talk to staff.
- Take entire teaching staff to curriculum centres, e.g. reading resource centres, units at universities, etc.
- Conduct liaison exercises with local secondary schools.
- Hire dance, drama or music companies to provide workshop experiences with staff.
- Arrange meetings with members of LEA, inspectors, etc.
- Pay tuition and examination fees for staff.
- Pay any residential charges for courses, and travelling expenses.
- Arrange residential conference in agreeable surroundings.

If the opportunities for in-service training are used properly, the staff of the school should be able to feel that they are learning and developing together, and this will assist the growth of the school as a whole.

PROBATIONERS

One specific training duty of the headteacher which should be taken in isolation from the rest of the development process is the induction of probationary teachers into the profession. This is usually a pleasure as well as a responsibility.

If we do take a probationer on to the staff, there are certain specific responsibilities which we must undertake in order to make her initial full-time experience of teaching enjoyable as well as useful.

The first steps in the introduction of a probationer are to:

- Ensure that the teacher has no accommodation or travel problems.
- Liaise with the local inspector responsible for the induction of

probationers, and ensure that the LEA's policy for new teachers is known and followed by all concerned.
- Take care to allocate the probationer a suitable class, in keeping with the type of training undertaken and containing no outstandingly difficult problems of control and discipline.
- Appoint a suitable experienced teacher on the staff to look after the probationer at all levels. Stress to the teacher and the probationer that what passes between them will not be divulged to the headteacher without the probationer's permission, so that the probationer can be frank with her mentor.
- Arrange for the probationer to have plenty of free time to sit in with other teachers and visit other schools.
- Arrange a timetable so that you visit the probationer at work in the classroom regularly.
- Arrange to meet the probationer regularly to discuss progress and any problems.
- Make sure that the entire staff is aware of the probationer's presence and the arrangements made to make her feel at home.
- Make sure that the probationer knows the details of the curriculum of the school and the standards of attainment expected of her class.
- Ensure that the probationer knows the safety and health regulations of the school.
- Ask a colleague to take the probationer round the school to familiarise her with the layout and the procedure for obtaining stock, etc.
- At the end of each term have a longer than usual talk with the probationer, letting her know how you think she is getting on.
- Liaise with the local inspector responsible for new teachers before making any written reports on the teacher.

When we appraise the probationary teacher at work in the school we should, within reasonable limits, apply the same standards and expect the same results from her as we do from the rest of our colleagues. Because of the particular position of the newly-qualified and inexperienced teacher, however, there are a number of specific qualities we should look for, in order to make sure that our new colleague is making the transition from the sheltered world of college to the hurly-burly of the classroom:

- Is she aware of the individual needs of her children?
- Is she able to translate the curriculum into actual lessons?
- Is she getting on with all colleagues, not just the teachers?

- Is she able to criticise herself and be aware of her mistakes?
- Is she coping with the administrative side of running a class?
- Does she seem willing to keep on learning?

In any discussions with a probationer, the head should endeavour to get her junior colleague to analyse her own performances as far as possible. We must each of us discover the wheel for ourselves, and conclusions we reach about our own competence make a greater impression than those imposed upon us by others.

The probationer should be asked to work out what she likes best and least about her work, where she thinks she is the most useful and least valuable to the school at the moment, and so on. We must take care to let our young teachers keep cheerful, by reassuring them constantly that they are settling in well, unless of course this is not the case!

One aspect of introducing a probationer is the possible effect on parents. Naturally we all want the best for our children. The arrival of a new teacher always causes some anxiety, and there could be more unease if the newcomer to the staff is rumoured to be a 'learner'.

In a case like this, assuming that the probationer is doing well, the children will usually report back favourably on their new teacher. It is also a good idea, in a low-key and unobtrusive fashion, for the head and the more experienced teachers to let their satisfaction with the new teacher be known in casual asides at PTA functions. Once it gets round that 'they're ever so pleased with Miss ... at the school.', it should allay any apprehensions.

THE DEPUTY HEADTEACHER

To be second-in-command of any organisation is not an easy role to fill. Unfortunately there have been too many examples of animosity between heads and their deputies. This is a shame, because when the head and the deputy do work in tandem, the efficiency of the school more than doubles. On the other hand, if the deputy is treated as just another teacher, then trouble usually follows.

If we have the opportunity to assist in the appointment of our own deputy, then we should favour a candidate whose skills and aptitudes are complementary to our own. If the head tends to be a detail person then the deputy should be the one with the ideas and vision, and vice versa. If the deputy is already at post when we arrive at a school, then a conscious effort must be made by the head to divide the leadership tasks in an equitable fashion.

As with everything else in the school, it is up to the head to set the tone of the relationship with the deputy and to delineate the latter's areas of responsibilities. An atmosphere of mutual trust and respect should be sought for. The deputy should be taken into the full professional confidence of the head, and the responsibilities given the deputy should be important and meaningful ones, not just odds and ends with which the head cannot be bothered.

It should be the aim of the headteacher to train a deputy who will be valuable to the school in his or her own right, who could take over the running of the school, who is able to act as a liaison officer between the head and the rest of the staff, and who is being trained to become a headteacher at some time in the future.

In order to carry out these duties the deputy will have to be taken through all the tasks of the head. It is essential that the deputy has time to carry out the extra duties provided and to gain experience of running the school. Wherever possible, the headteacher should timetable the deputy so that he or she has at least one block of free periods a week, perhaps for two hours, to be in charge of the school, meeting visitors, dealing with telephone calls, etc., while the head takes the deputy's class.

Above all, the headteacher must show by word and deed that she regards the deputy as an integral part of the school's management team. If the headteacher's attitude towards the deputy is made plain, then colleagues, parents, governors and other visitors will adopt the same policy.

Non-teaching staff

The ancillary helpers, secretaries, caretakers, and other non-teaching members of the staff all have a vital part to play in the smooth-running of the school and should be treated as the valuable and valued members of the school family that they are. Their worth to the school should be made evident to them and they should be granted the full privileges entitled to them as colleagues. They should be welcomed into the staff room and included in all social functions. In a happy, well-integrated school the ancillaries will usually be doing far more than their allocated jobs; many of them will also be taking clubs, accompanying teachers and children on trips, and helping with games and concerts.

The description of the ancillary helpers' work will depend very much on the specific needs of the school, as will that of the midday meals supervisors, but the duties of the secretary and caretaker can be outlined.

THE SCHOOL SECRETARY

The job description for school secretary might include:

- Generally helping the headteacher with the organisation and administration of the school.
- Acting as telephonist and receptionist.
- Completing any necessary forms and returns for the headteacher's signature.
- Taking charge of typing, filing and duplicating duties.
- Maintaining all necessary records of staff and children.
- Helping the headteacher with capitation allowances and school fund.
- Assisting with salary and wages queries.
- Maintaining first-aid stock in school.
- Keeping records of attendance of children.
- Distribution and checking of attendance registers.
- Any other duties which may reasonably be asked by the headteacher.

THE CARETAKER

With the reorganisation of services and the introduction of private cleaning firms to take over in schools, it could be that before long in many areas the school caretaker will be an extinct species. This will be a great pity. For the moment a typical job description for a school caretaker might be:

- Being responsible to the headteacher for the upkeep and safety of the building and fabric of the school.
- Locking and unlocking the premises at the beginning and end of the school day.
- Maintaining the heating and lighting of the school at an acceptable level.
- Cleaning and dusting the premises.
- Picking up and sacking litter in playground.
- Checking on security of premises.
- Supervising school during evening and weekend letting of premises.
- Reporting major defects to building or plant, and liaising with firms employed to repair them.
- Effecting minor running repairs where possible.
- Any other duties which may reasonably be requested by the headteacher.

Dismissal of teachers

The process of dismissal can take different forms. Whatever form it takes, dismissal is a traumatic experience for all concerned. Quite literally, it will often end in tears. No one who has been involved in the process in any capacity ever wants to repeat it. Fortunately, most headteachers can complete their careers without ever being involved in a dismissal, but sometimes it must be considered.

As headteachers we have both a legal and a moral obligation to explore every avenue before contemplating any of our colleagues for dismissal. Not only is it very difficult to remove a teacher from a school, except for gross moral turpitude or shattering neglect or duty, but it implies that there has been something wrong with our processes of selection or in-service training.

If the teacher concerned is an experienced one and can point out that no other head has implied that she is not fit to teach, then the problem becomes even greater. For the probationer teacher who does not reach the required standard, there are definite steps which may be taken with the LEA and its inspectors before the probationer is granted fully qualified teacher status. This might mean an extension of the probationary period.

For an experienced teacher who does not seem able to improve, despite counselling, training, advice and encouragement, a change of class or even of school, it may be concluded with great reluctance that dismissal procedures may have to be considered.

First the head should consult her governors and the LEA, to discover the local policy on teacher dismissal. The inspectors and perhaps the governors will want to talk to the teacher concerned and see the teacher at work. It may well be that the governors will already have heard something about the teacher if her actions are causing concern among the parents.

Before any dismissal procedure begins the teacher should receive an informal warning from the head about her shortcomings. If there is no improvement, an official formal oral warning must be issued. This should be given by the head to the teacher in the presence of a senior member of the teaching staff. The teacher must be told in advance the reason for the meeting and allowed to bring a friend or union representative. Details should be placed in the teacher's file and the governors and LEA informed.

If the teacher still gives cause for anxiety, the next step is to issue a formal written warning, after a meeting involving the Chief Education Officer or a delegated officer, the head teacher and the teacher. The written warning will state that unless there is an improvement within a stipulated time, a final written warning may be issued, or dismissal procedures set up by the LEA.

Five reasons for dismissal are cited in 1978 Employment Protection (Consolidation) Act:

1. Capability and qualifications: assessed by reference to skill, aptitude, health or any other physical or mental quality, and academic degrees or diplomas relevant to the employee's position.
2. Conduct: assessed by behaviour and discipline at work and, where this might affect work, by conduct outside working environment.
3. Redundancy: this may be involved if place of work closes, or if employer no longer needs the kind of service performed by employee.
4. Statutory restriction: may be employed if worker cannot do job without breaking law, e.g. a driver who loses his driving licence.
5. Some other substantial reason.

The headteacher must take care not to provide an employee with grounds for a claim for wrongful dismissal. All local and national guidelines covering the situation must be known, and the teacher concerned made aware of them. Above all, the headteacher must be scrupulously fair. If a teacher has shown herself to be unsatisfactory, then the head must be able to produce evidence of this fact. A written record of any lapses on the part of the teacher should be kept, and the head should discuss the matter with her deputy or a senior colleague.

Staff facts

ASSAULTS ON STAFF

If a member of the staff is assaulted by a parent, the first duty of the headteacher is to ensure the safety and welfare of the colleague who has been attacked. Assistance and comfort should be given to the colleague, and arrangements made to see that she is taken home.

The parent concerned should then be asked to leave the school premises. As a rule, after an explosion of this nature, the guilty party is only too willing to put some distance between himself and the school. Under no circumstances should the headteacher attempt to use force to eject the parent. If he refuses to leave the police should be called. As a rule they will take no action in such a case, but they will see that the parent leaves the premises, as he may now be deemed to be trespassing. When the situation has been defused, any children present have been reassured and put in the charge of a teacher, the next steps may be taken.

This will mean taking a statement from the teacher concerned and accounts from any adult witnesses. The LEA should then be informed of the situation and asked to take appropriate action. The teacher concerned could be advised to consult her union, and the headteacher would be well advised to consult her own professional association.

If the LEA agrees to prosecute the parent through the Criminal Injuries Compensation Board, the parent concerned will have to give evidence. If he is found guilty of assault, the Board is empowered to meet claims for compensation in excess of £400, if the parent is incapable of paying. If the LEA is unwilling to take this action, the headteacher and the teacher involved should consult the legal departments of their respective unions and take their advice. The unions, if they think the case merits it, may take action on behalf of the school.

In the meantime, when matters have cooled down, the headteacher should contact the LEA and enquire what steps it is prepared to take to ensure that teachers and other members of the staff are not subjected to further attacks.

The head could then decide to see the parent concerned and set down some guidelines for the future. Plainly it would not be fair to persecute a child because the parent was aggressive, but the parent should be informed that he must undertake to support the disciplinary policy of the school. The headteacher should then inform the parent that he may only visit the school after making an appointment, and that he is not to attempt to contact the teacher involved in the assault. The headteacher must deal with the parent from now on. If the parent has been prosecuted for the assault he will probably agree to these terms. Should he refuse to do so, however, the LEA should be contacted and asked as a matter of urgency to support the headteacher and the staff.

If a child assaults a member of the staff, the parents should be brought in at once and asked to make sure that it never happens again. Should the offence be repeated, the governors should consider excluding the child for a fixed period, after the Educational Psychologist has had an opportunity to assist.

If, after all these efforts, the child remains aggressive, the LEA should be informed that in the opinion of the school the child is disruptive and should be educated elsewhere. Teachers should be advised by the head not to refuse to teach the child but to be patient while events take their course. As a rule, some form of special education will then be devised by the LEA.

Summary

It is part of the headteacher's duty to recruit, train, motivate, encourage, appraise and where necessary discipline staff members. The ethos of a school depends upon happy, hard-working, fulfilled people working as a team.

It is worth bearing in mind what staff require of a job:

- The opportunity to do interesting and creative work.
- A considerable amount of responsibility.
- A good working environment.
- Good career prospects.
- Suitable financial rewards.
- Pride in working for a prestigious organisation.
- Security.
- A feeling of being in control of one's action.
- Recognition.

Self-assessment

The headteacher should work out just how much she knows about her staff. Are they all distinct, separate human beings to her? Does she know their strengths and weaknesses and aspirations? Does she know how to deal with each one in the most appropriate manner? Is she capable of welding them into a team?

QUESTIONNAIRE

Recruitment of teachers
Draw up a suitable general description of your school to accompany the specific job requirements if a teacher should leave. In this general section include details of the area, school, curriculum, out-of-school activities and home–school links.

Deployment of staff
In which particular areas do you feel that your school is still weak in terms of qualified staff? Do you need more musicians, scientists, games coaches, domestic science specialists? Make a list of the additional strengths you feel you still need in the classroom, and out of it.

---

continued overleaf

Communication with staff
If you are not already doing so, draw up a staff newsletter to circulate to your colleagues, keeping them up-to-date. Analyse the general reactions to this and decide if it is an exercise worth persevering with.

Appraisal and training
Drawn up (a) a self-appraisal form for your teachers to fill in, and (b) a self-appraisal form for yourself. Work out and write down a staff training programme to occupy the five Baker days in a school year. On which particular areas do you feel your school needs to concentrate?

Non-teaching staff
Draw up or revise detailed job descriptions for your school secretary, caretaker, dinner supervisors and any ancillaries.

Chapter 5
The Parents

Even the most progressive and open of headteachers can be a little apprehensive about contact with parents, but the majority of headteachers of course welcome the assistance of parents in tthe running of their schools, accepting that this should be an integral feature of today's education system.

There are many ways in which we can involve parents in the lives of our schools. The first and most important is to gain their trust by making sure that every child receives the very best care and attention that we are capable of bestowing. Only when this has been accomplished can we hope to start weaving that intricate mesh of relationships necessary if we are to make our schools answerable to the parents of our children, and at the same time include as many of them as possible in the day-to-day administration and organisation, as well as taking on board their advice about the curriculum.

Drawing parents into the school

Parents can be attracted towards a school even before their children join a reception class. The headteacher should take advantage of the various pre-school organisations in the area – mother-and-toddler groups, play schools, Young Mothers' clubs, and so on. By making the rounds of these and giving a talk on how children may be prepared for their first year at school, the head may develop a reputation for willingness to talk to parents. If this talk on getting pre-school children ready for school is illustrated by slides of activities at the school, then a public relations exercise may be performed at the same time. The parents who are encouraged to enrol their children at the school may then be persuaded to take a full part in school activities at a later date. Topics covered in this talk could be divided into two:

Physical preparation to see that your child can:

- use the lavatory
- make its wants known
- know its name and address
- dress and undress
- expect to clear up toys, etc.
- sit and listen to a story
- use eating and drinking utensils
- hold a pencil and crayon
- manipulate door-knobs, latches, taps, etc.

Mental preparations to help your child to learn how to:

- get along with other children
- get along without the presence of parents
- become accustomed to handling books
- look forward to being read to, etc.

PARENTAL INVOLVEMENT
When parents have young children at the school, it is up to the headteacher to provide a climate in which mothers and fathers know that they will not be rebuffed if they offer to help. The school brochure should provide a list of activities with which the parents can help in and out of school hours. They might include:

- Helping with clubs and activities
- Listening to children read
- Accompanying staff and children on journeys
- Repairing and sorting school library books
- Washing and ironing school sports gear
- Making costumes and scenery for school plays
- Selling Christmas cards for school fund
- Providing transport, if properly insured, for school outings
- Helping to dress and undress infants for PE and games
- General classroom help – setting up and clearing up
- Helping with repairs and renovations at school
- Getting involved in school building projects, e.g. building a swimming pool
- Helping with holiday play schemes
- Helping with creche to enable other parents to leave children and attend school functions, etc.

LOOKING AFTER PARENT VISITORS
Whether parents are involved in helping teachers prepare for an art lesson in the classroom, refereeing a school football match, or advising the head on an aspect of the curriculum about which they have special knowledge, it is important that they be made to feel welcome and useful. They should also be informed exactly what is expected of them and not left to drift around looking for things to do. Their surroundings should be as comfortable as possible when they do visit the school to help. Details of arrangements should be set out in a letter to parent helpers.

Arrangements should be made for tea or coffee breaks for parent-helpers, either in the staffroom, or with the head in her office, or, if space permits, in a special room set aside for parents. If there is such a room it should be equipped with comfortable chairs, a kettle, cups and saucers, and a notice board which could contain a list of tasks to be done – books to be covered, posters to be drawn, etc.

Communicating with parents

In addition to taking every opportunity to talk to parents who come in to the school, there are a number of other forms of communication which strengthen contacts. These include the school brochure, newsletters, reports, home–school diaries, anthologies of children's work, and the telephone.

THE SCHOOL BROCHURE
All parents should be given a copy of the school brochure and any subsequent amendments should be sent to parents as they are made. The brochure should have a cheerful and striking cover, with the full name of the school. It should be available at least six weeks before the term to which any information in it applies, and should be used to advertise and 'sell' the school, as well as providing information about it.

The school brochure should cover the following topics:

- Brief introduction from chairman of governors
- Name, address, telephone number and location of school, with sketch map
- Type of school – county, aided, controlled, etc.
- Any church contacts and affiliations
- Arrangements for pre-school children to visit school
- Ages at which children are admitted

- Whether the governors intend reserving a limited number of places for anticipated numbers of children living in area
- Details of opening and closing times, and when children are allowed in playground in morning before school
- If produced every year, the dates of the terms
- Names of headteacher, teachers, non-teaching staff
- Names and addresses of governors and caretaker
- Details of administration of school:
    Behaviour and discipline
    Grouping of children in classes
    How parents should notify absence of children
    Policy of school if child is taken ill or has accident
    School meals provision and supervision
    Any policy for sex education
    Out-of-school activities, including open evenings
    Fund-raising activities
    Policy for children with special needs
- Details of PTA, parent-governors, ways in which parents can help school
- Any school uniform or special clothing needs
- Details of curriculum, including provision for different age-groups, RE policy, etc.
- Any homework or home–school reading schemes
- Insurance provision
- Any recent sporting or dramatic successes, etc.
- Responsibilities of pupils for looking after personal possessions
- Any other relevant information
- Closing letter from chairman of PTA

NEWSLETTERS

Newsletters should be used for keeping parents in touch with what is going on and what is going to happen at the school.

Many school newsletters are still produced by stencils and then duplicated, but in recent years more and more primary schools have been using the more attractive and eye-catching opportunities of a computer. A commercially-produced disc will provide the format of a newspaper layout for the program, with headlines of different-sized typefaces. The user can keyboard in the spaces under the headlines, and operate a mouse to draw illustrations for the newsletter. Many programs can be operated by children as well as adults. Some programs will give more elaborate opportunities for design than others, and will depend upon the type of computer available in the school. Among those programs in common use are NewSPAper, Wordcraft, Typesetter, Fleet Street Editor, and Front Page, but there are others.

There are different ways of producing enough copies of the completed newsletter, depending upon the type of technology available. The layout may be transferred to a heat-copier or a photo-copier.

The main problem with the production of school newsletters lies in making sure that they actually get home. Children seem to make a practice of depositing them in the bottom of lunch-boxes and school bags, where they may lie undetected for months at a time, or even losing them altogether on the way home. Apart from recommending that parents carry out regular body-checks and searches on the persons of their children, there are a number of ways of reducing the chances of our newsletters going astray:

- Try to issue them at the same time each month, e.g. in the first week of each calendar month, so that parents are expecting them.
- Issue a tear-off strip at the bottom of each newsletter, so that parents can sign it to acknowledge receipt and return it via their child.
- Number newsletters consecutively by issue over the course of a year, so that parents may notice if they do not receive copies in sequence.
- Put a copy of each newsletter on the school notice board, so that word-of-mouth information may circulate among parents that the latest issue should be on the way home.

REPORTS

Many primary schools now sent home reports on the work of children. These are basically the work of class teachers, who give the results of tests taken, reading ages, comments on behaviour, and so on. The headteacher should also reserve a space on each report to make an overall comment on the work and conduct of the child. The head's contribution should be more open-ended and informal than that of the class teacher, and should endeavour to produce in a few words a profile of each child and its place in the class and the school. Sporting and aesthetic achievements and interests should be included, as well as academic results. It is important to include a remark or two which will show that the headteacher does in fact know the child and is interested in his progress.

HOME-SCHOOL DIARIES

In special circumstances the headteacher should also consider the use of home-school diaries with some children. These are particularly effective in the case of children with learning or behavioural problems, and provide an area in which the head may take some of the load off a busy class teacher.

If a parent has been brought in to the school to discuss the behaviour or lack of progress of a particular child, an arrangement could be made by which, at the end of every week, the head sends home by the child concerned an account of that child's work and behaviour over the past five days and suggestions for work to be done over the weekend. The parent could sign the book and add any appropriate remarks.

Letters

Occasionally a headteacher will have to write an individual letter to a parent about a breach of school rules or any other occurrence which has caused concern in the case of a child. Every effort should be made to avoid 'headspeak' in such letters. Getting a letter from the school is a significant and often dismaying event in such households. If this letter should be couched in pompous or hectoring terms it may well prove to be counter-productive. The head should take care to see that any such communications are written in a friendly, fair and straightforward manner.

Taking work home

Not all communications home need to be or should be complaints or requests for money. Parents are entitled to see the products of their children's education. Boys and girls should be encouraged to take their work home as much as possible. At the end of term, when wall displays are taken down, the individual children should be given their contributions to take home. In the same manner, any project work, anthologies and completed exercise books should also be sent home to the parents. This is usually a matter for the class teacher, not the head.

The telephone

The telephone should be used by the headteacher to convey good news as well as bad. If a child has done an outstanding piece of work and the head knows that the parent is home during the day, then a quick telephone call to that parent should reinforce the reputation of the school as a friendly and caring place.

When the telephone has to be used to inform a parent that a child is giving cause for concern, the matter should be put to the parent in an open, straightforward and *brief* manner, and the parent asked to come along to the school to discuss the matter.

Person-to-person contact

In the last analysis, it is the person-to-person contact between the headteacher and a parent which has the most effect. We should take great care with this important aspect of our managerial and public relations work. Should we develop a reputation among parents for arrogance, flippancy or any other off-putting characteristics, it will swiftly undo any good we might be accomplishing in the school.

We should remember that some parents will have unhappy memories of their own school days and may be nervous or uneasy about coming to see the headteacher. To others we may represent 'authority' and as such present a threat. This trepidation may occasionally reveal itself in apparent displays of animosity or discourtesy. For the sake of the children concerned we should do our best to control and guide such interviews, so that some good may emerge from them. These points may be worth following:

- Be welcoming, friendly and dignified.
- Offer the parents coffee. Sit at the same side of the desk with them. Do not interview them from behind a desk.
- Take care to have all relevant records of the child concerned with you. Study them in advance and let the parents know from your attitude and comments that you are aware of any problems with the child.
- Try to work out if there is something else worrying the parent other than the ostensible cause for the visit. Sometimes the real concern of the parents will not at first reveal itself.
- Listen quietly and sympathetically to everything the parent has to say. Do not interrupt and do not argue.
- Do not discuss any member of the staff or the parent of any other child, unless those people are present at the interview. Say that you will take up any points raised with that teacher or parent, and get back to the complainant.
- Develop and deserve a reputation for being discreet, so that parents will confide in you.
- As far as possible, without betraying confidences, try to give parents a full and frank explanation of any event which is bothering them.
- If, after the interview, the parent is still not satisfied, offer to call another meeting with the Chairman of the Governors and other governors present.

The vast majority of interviews between headteachers and parents are amicable and productive affairs. On the rare occasion that a parent comes to the school spoiling for a confrontation, we have a professional duty to remain calm and objective about the whole matter. Above all, we should endeavour to impress upon the parent that it is the child who matters, not the wounded pride of the parent.

It must be said, however, that there will be times when all headteachers will come into contact with a parent whose attitude can only be termed unreasonable. Anyone who has been in the presence of an enraged mother or father and suffered a tirade of insults or imprecations will know what a sickening experience it is. It is not what any of us came into the job for.

The only answer to such behaviour, if the parent refuses to listen to reason, is to suggest that he or she takes the matter to a higher authority, and then quietly walk away before the confrontation escalates. It will not be much of a consolation, but if we are sure that we have done everything within our power to ameliorate the situation without success, we can at least take comfort in the fact that it is impossible to please all the people all the time.

Involving parents in the education of their children

There is an increasing trend towards involving parents in the education of their children. This pattern is to be welcomed, as it helps to foster closer links between home and school.

The first step that a headteacher should take in this direction is to let parents know what is going on at the school. This can be done in the school brochure and newsletters, but the most effective method seems to be that of holding a curriculum meeting at the school, where the head informs the parents of what the school is trying to do in the core and foundation subjects and other areas of the curriculum.

A good time to hold such a meeting is as an adjunct to the compulsory governor-parent meeting at which the governors' report is discussed every year, especially if this is held towards the beginning of the school year, perhaps in October or November. The headteacher could give her talk after the main meeting is over, perhaps following a coffee break. It will add interest to the proceedings if the talk is illustrated with slides taken in the classrooms, or even with a specially prepared home-made video production.

The talk could cover:

- How the classes are organised
- The aims of the different classes in the core and foundation subjects
- Methods of teaching the core and foundation subjects

- How the children are tested and assessed
- How parents can help their children

If time permits, the parents could divide up into groups, each group being taken by a different teacher who will demonstrate how individual classes or subjects are taught, giving the parents a chance to join in. There could then be follow-up evenings at which different teachers, assisted by the head, elaborate on the methods of teaching and aims of the individual classes.

If the teachers agree and there is enough space in the school, a further follow-up activity could be an Open Day, when parents come in to the school to watch the children being taught.

LOG BOOKS

For those heads who are really committed to the idea of keeping parents in touch with what their children are accomplishing at school, the older boys and girls could take home once a month a log book in which they have outlined the work they have been doing over the previous four weeks. This will usually entail the head in taking the class for the last afternoon of the month, helping the children fill in the blank spaces in their duplicated log books. These headings give an outline of a log book:

- Mathematics
  What sorts of mathematics have you done? Write down some examples.
- Creative writing
  What stories have you been writing? Say something about them.
- Reading
  Which books, stories and poems have you read? Write down the titles. Which did you like best? Say something about them
- Handwriting
  Copy down six lines of your own choice as an example of your handwriting
- Science
  Write down something about the science you have done this month
- Social studies
  Write a few lines about the history, geography and local studies you have been looking at
- Religious and moral education
  What have you been discussing in religious education?

- Design and technology
  What sort of plans have you drawn and how have you put them into practice?

If the headteacher does spend one afternoon a month with the older children, helping them to fill in their log books in this way, it will enable the children to revise the work they have been doing and enable the head to check that the syllabus is being followed.

HOME-READING SCHEMES
The consensus of opinion so far seems to be that the most effective form of home–school learning is for teachers to recruit parents to help their own children with their reading at home, and to set up some sort of recording and testing process in order to monitor results.

One of the earliest pieces of research carried out in this area was by Professor J. Tizard and his colleagues at London University, who studied a number of seven-year-old children from working-class homes in Barking. It was found that one factor stood out in determining children's reading progress at school – whether or not their parents were helping the children with reading at home. Parental assistance made all the difference between children being above average and below average readers for their ages.

Tizard and his team went on to look at what would happen if schools set out systematically and deliberately to enlist parents as teachers of reading. In Haringey, a group of schools agreed to participate in a scheme in which children would take home their books every second night. The results of tests showed that after two years, more than half of these children had reading ages above the average for their ages.

Following the encouragement provided by the Plowden Report, a number of boroughs and individual schools began involving parents in their children's reading activities. In 1981, *The Times Educational Supplement* commented:

> Involving all parents directly in teaching their own children is quite different from the rather vague messages and practice on home–school links that are delivered in many primaries.

As a result of these experiments, a number of schools were able to provide concrete advice to others wishing to involve parents in reading schemes. In Hackney, an organisation of teachers, parents and inspectors called PACT published an account called *Reading Partnerships in Hackney* (1984). Parents were asked to listen to their children reading at home and record what they and their children thought about the books and any problems that arose. The scheme emphasised the importance of the three ps: praise, patience and pleasure.

The home reading scheme operated by Belfield Primary School in Rochdale was described by Peter Wilby in the *Sunday Times* in 1981. Parents were given a simple list of do's and don'ts to help them listen to their children reading at home.

Do
- Let your child sit close to you
- Talk about the pictures first
- Read the page to him first
- Smooth out the difficulties by telling him words he doesn't know.

Don't
- Make reading an unpleasant task
- Threaten to tell the teacher if he doesn't do it
- Make him think he is in competition with others
- Have the television on.

In Sheffield, the staff of Fox Hill First School launched a similar home-school reading plan in 1980, and four years later summarised their achievements in the pamphlet *Have You a Minute?: The Fox Hill Reading Project*, by Hilda Smith and Margaret Marsh:

> We found that parents became very involved with the project ... Reading regularly at home to their parents has improved the level of reading of the children. Although we have no tests to substantiate this statement, we do not feel it is necessary to have 'proof' of our success, we know the project is successful.

Other useful publications on how to use home reading schemes are *Children and Parents Enjoying Reading*, by Peter Branston and Mark Provis (Hodder and Stoughton, 1989), and *Shared Reading in Practice*, by Chris Davis and Rosemary Stubbs, (Open University Press, 1989). Among the commercially produced series which could be considered as the basis for home reading schemes, are *Hello Reading* (Puffin), *The Parent and Child Programme* (Octopus), *Help Your Children Storybooks* (Collins), and *Read Together* (Piccolo).

The best way for a headteacher to initiate a home-school reading project is by calling interested parents and teachers to a meeting. Some schools have found that calling small groups of parents in to discuss the proposed scheme is more successful than assembling large numbers for the same reason.

There are several stages in launching a home–school reading scheme.

1. Help the parents understand how best to help their children with reading. This may be done by a straight talk, the use of a video, or by role-play where the staff, parents and children demonstrate correct and incorrect ways of reading with children.
2. Hand out prepared booklets summarising what has been said at the meeting.
3. Distribute record sheets which allow teachers and parents to make comments about children's reading. These could have the headings: book, date, pages read, time spent, parents' comments, teacher's comments
4. Stress that the home-reading sessions should not last more than ten minutes at a time. The number of sessions should be left to the judgement of the parents.
5. The headteacher, wherever possible, should take the burden off the class teacher by checking the record cards as they come in every week, and add the teacher's comment before sending them back.

Obviously, before such a scheme can be launched, the head must make sure that there are sufficient suitable reading books in the school. If there are not, fund-raising activities will have to be considered.

HOMEWORK

Some schools operate more wide-ranging homework schemes than just home–school reading projects. In 1985, the DES consultative document *Homework* set out the following possible objectives for any homework scheme:

- To encourage pupils to develop the practice of independent study.
- To develop perseverance and self-discipline.
- To allow practice, where it is needed, of skills learned in the classroom.
- To permit more ground to be covered and more rapid progress to be made.
- To enable classwork to concentrate on those activities requiring the teacher's presence.
- To open up areas of study and to make possible the use of materials and sources of information that are not accessible in the classroom.
- To involve parents (and other adults) in pupils' work.

Another DES publication, *Education Observed 4: Homework* (1987), defines four main categories of homework:

1. Homework which is closely integrated with and reinforces classwork and has clear curricular objectives. This includes practice in the skills of literacy and numeracy.
2. Homework which exploits materials and resources in the environment outside school. Children and their parents may make studies of local natural resources to link with school work.
3. Homework which encourages independence, research or initiative. This may include such open-ended tasks as looking things up in books at home or in the library.
4. Homework which involves pupils in working with parents and other adults in the community. Here children and parents can work together on such things as cookery or handicrafts.

If the school does decide to adopt a system of homework, children should not be asked to spend too much time on it, perhaps a maximum of half an hour a night for two or three nights a week.

Parents as policy makers

On an increasing scale parents are influencing curricular policies, and if schools are to progress it will have to be with parental approval.

In some areas parents are contributing to the school curriculum, and this will certainly increase. Most parents, however, will prefer to be assured that their own children are being well taught and properly looked after.

The headteacher should attempt to keep her finger on the pulse of local opinion by meeting as many different parents as possible. At first most of them will not have many clear-cut ideas about the direction the school should take, except insofar as their own children are concerned. They will want them to be happy, secure and well-organised against a background of sound pastoral care, good home–school relationships and communications and a wide variety of interesting out-of-school activities. They will want the school to be known locally as a 'good' one. If all these conditions are met, most parents will be content to leave the actual organisation of the school to the head and governors.

Where parents will insist on a voice is in expressing their opinion as to whether or not the school is meeting their requirements. A head who knows what is going on will soon pick up any murmurings from the school gates or meetings in the local supermarket, and the governors and the PTA members should also let the headteacher know whether or not things are going smoothly.

QUESTIONNAIRES
In order to get a more balanced view of the opinions of the parents about the running of the school, the headteacher could well consider sending out a suitably-worded questionnaire. Some heads, believing in leaving well enough alone, may consider this to be tempting providence. It is true that such a method can only be used sparingly, perhaps once every seven years, i.e. once in the career of a child at the school, but it can certainly provide a headteacher with an interesting insight into how others regard the school. The questionnaire, if it is to be of any use, should concern itself with practical aspects of the school to which parents can give an objective opinion. Such open-ended questions as 'What do you think of the school?' should be avoided, as they are almost impossible to answer in a few words. In addition, too many less than enthusiastic responses might depress the head and staff too much!

A letter along these lines might be sent to parents:

Dear Parents,
Would you please let us know what you think of our efforts to keep you in touch with what is going on at the school?

There are a number of possible answers below. Would you please tick the answers that represent what you feel. If you would then return the questionnaire to the school it would be much appreciated. Do not sign the form. All returned questionnaires will be placed in a box in the school hall. There will be complete anonymity.

Open evenings
a) I am satisfied with the present arrangement and consider it sufficient to keep me in touch with my child's progress
b) I do not like the present system and would like it replaced by another

General access to teachers
a) I like the present informal system of coming in to see teachers if I have a problem
b) I would prefer a more organised system of making appointments to see teachers

Newsletter
a) I like the present informal monthly newsletter
b) I would prefer a more detailed newsletter
c) I don't like newsletters, feeling they serve no useful purpose

Helping in school
a) I feel that parents should only come in to help when there is a specific need
b) I feel that parents should be used much more to help in the school

Homework
a) There should only be homework if the teacher and parent feel that it is necessary for individual children
b) There should be regular, organised homework for all children
c) I think there should be no homework as the children work hard enough at school

Headteachers should decide which particular aspects of school life they need more feedback on before sending out questionnaires. We should not be disappointed if the response seems poor. Anything over a 50% return of completed questionnaires is good.

HELP WITH CURRICULUM
If there are parents with particular skills or interests who can be enlisted to assist in the drawing up of the curriculum, they should be approached at once, although of course the headteacher will have to have the final say in the design of the school's curriculum. Parents, say, with the ability to help with local studies, domestic science, games, etc. can make a significant contribution during the initial planning stage of the school's curriculum.

Open evenings

All primary schools should have an open evening every term when parents can come in to the school and look at their children's work and discuss this with the teachers. The teachers should accept this duty as a part of their directed time.

Open evenings can be a very pleasant and productive part of the school calendar, giving as it does teachers the opportunity to see most of the parents, and the parents the opportunity to look at the school. It is only sensible to see that the display on the walls is attractive and up-to-date, as this may well be the only time that some parents enter the building.

Some headteachers take the opportunity to give a talk to the assembled parents, but most prefer to see a steady flow of parents coming and going throughout the designated period.

A few schools ask the parents to come during school hours and talk to the teachers while the children get on with their work. Although this does give the parents the chance to see the school in operation, few teachers like this method and it does put a great deal of strain on their tempers, stamina and fund of good-will towards the head. Another method is to send the children home early one afternoon and bring the parents in then, but generally the middle period of the evening is favoured for these occasions.

For a class containing about thirty children, a three-hour span is usually enough for an open evening. The period 6 pm until 9 pm, or 7 pm until 10 pm will give parents who work a chance to get to the school during the evening.

Some schools operate appointment systems, with each parent in a class given a separate ten-minute segment in which to see the teacher. Many prefer to ask parents to come in at any time within the stipulated period that suits them, and write their names on the blackboard as they arrive so that the teacher may see them in the order of their arrival.

Parents should have been asked in advance not to spend more than seven or eight minutes with the teacher, so that all the parents may be seen in the time allotted. Parents with special problems which might take longer to resolve should be encouraged to make appointments to see the class teacher or the head before or after school one day.

The headteacher should try to station herself inside the school entrance, to greet parents as they arrive. She should be prepared to see individual parents privately if they request this. On the whole, however, open evenings are more occasions for the class teacher than the head.

Parent-teacher associations

PTAs have not been in existence in many schools for very long, relatively speaking, yet in a short time they have been through a number of changes. At first they were treated with suspicion by most heads and teachers, but as PTAs became recognised as helpful and supportive institutions, the widespread recognition that all parents should be involved in the life of the school made some heads wonder if home–school associations might prove superfluous.

This is not true. Some parents will have more time and energy than others to help in the life of a school and will welcome an outlet for their goodwill. A PTA provides a cadre of willing and available parents which can do much to enhance the development of a school.

A parent-teacher association provides an extra link between home and school, gives a school a number of 'willing horses' for most out-of-school activities, and helps keep the staff in touch with what the parents are thinking. Care should be taken to involve the PTA in educational aspects of the school's life; its members should not be regarded as mere fund-raisers. The head should also ensure that parents who do not belong to the committee do not regard the PTA as an elitist organisation whose members get special consideration from the school.

MAKING THE MOST OF A PTA
The PTA should have a constitution which states its aims and intentions, and stresses that the final voice in educational matters must be that of the head and governors. An excellent draft constitution and much valuable advice on forming and running a PTA may be obtained from the National Confederation of Parent Teacher Associations, 43 Stonebridge Road, Northfleet, Gravesend, Kent, DA11 9DS.

The headteacher should not be chairman of the PTA, in case any representations have to be made to the LEA. A parent who is not employed by the authority will have much more room in which to manoeuvre.

Minutes of the PTA meetings should be circulated to all parents, governors and teachers, perhaps as an appendage to the school newsletter.

It is usually advisable not to limit the size of the PTA Committee. Let all who wish to join come on to it. There is little chance of the body being swamped, and every extra pair of hands will help.

Try to set every PTA Committee member the task of linking up with at least ten other parents outside the committee, bringing them in to help the school, asking them to buy tickets for fund-raising activities, etc.

Ask the PTA to raise funds for certain specific activities – paying for the children's Christmas parties and entertainments every year, etc.

Dealing with a bereavement

It may sometimes be the sad duty of a headteacher to have to deal with the aftermath of a bereavement. Each case will require a different response. Many heads, it must be said, would question the ethos of the school and the commitment of the staff if anyone had to consult a piece of paper in order to deal with a death connected with the school. Nevertheless, if a headteacher has not had this experience she might care to draw up general guidelines, or at least consider what her reaction might be to such a situation.

DEATH OF A CHILD
If a pupil at the school dies after an accident or illness, the effect on the

whole school may be considerable. The whole staff should be alerted and asked to look for reactions in other children.

It will help if the headteacher has already included a study of death and its implications in a religious context as a part of the RE syllabus. If a school pet should die, the opportunity could be taken to deal with the subject of death, and what the major religions believe about it, at a school assembly.

The headteacher should of course write to the bereaved parents and attend the funeral. It may be that one of the parents may come in to the school. If this happens, the head should be sympathetic and understanding. If the parent broaches the subject and obviously wants to talk about the child, then the head should comply with this request.

Other children in the school, and particularly in the dead child's class, should be watched closely for signs of grief or distress. Some of them may experience feelings of guilt, because they are alive and their friend is not. The class teacher may be able to deal with individual cases, if not the head should talk quietly to the children concerned. Any close friends of the deceased child should be watched at playtime, allowed to be alone for a time and then eventually persuaded to join in with the other boys and girls at play.

Death of a parent

Children react to the death of close relations in different ways. If a parent or a close relative dies the child will almost certainly be away from school for a time. When he returns he should be watched carefully but allowed to come to terms with his grief in his own way.

If the surviving parent should visit the school, the head should again deal with the situation as she thinks advisable. Some parents may not know many other people in the area. The headteacher may be one of the few people the parent will be willing to approach to share the grief.

On a practical level, if the parent's financial circumstances have changed, the head, quietly and efficiently, should be able to provide details of how to obtain free school meals for the child, and whom to approach for help with travelling arrangements to the school from home. The aid of another parent may be enlisted to bring the child to school in the morning with her own children if the surviving parent is at work.

Parental facts

Parental duties
It is the duty of a parent to see that any child registered at a school

attends that school regularly. Acceptable reasons for absence include:

- Illness or unavoidable cause concerning the child himself, not the parents
- Days of religious observance by any group to which the parents belong
- If the parent can prove that the school attended by their children is too far from their home for the children to walk (providing that the LEA has made no effort to find a school closer to the child's home or supply transport for the child)
- Holidays taken by the parents in school time, as long as the child is not normally away for more than two weeks.

If a headteacher feels that a child's lack of attendance is sufficient to cause concern, the LEA should be informed. Usually poor attenders are nominated on a regular return submitted by the school to the authority. In the first instance the aid of the education welfare officer will be enlisted. If the parents insist on keeping the children off school, or if they condone their children's truancy, the LEA will probably issue a school attendance order to them. The parents will then be liable to a fine of up to £200 in the first instance.

If the parents can prove to the satisfaction of the LEA, or on appeal to the satisfaction of the Secretary of State, that a child is receiving a suitable education at home, then the child may be allowed to continue this home education.

RIGHTS OF ACCESS

If parents are separated, both parents have the right to see their children unless there is a court order in existence forbidding access to one of the parents. If a headteacher knows of the existence of such an order, she may refuse to allow the parent in question to see the child at school. If the headteacher is unaware of the existence of such an order from the court, then she must allow the parent full normal rights of access to the child, which will include taking the child home.

Provision should be made by the school during the time that a child is enrolled at the school for the custodial parent to inform the school of any court order banning the other parent from making contact with the child, or providing only limited access. If a parent arrives at the school and the headteacher knows that he or she is denied access to a child, then the head should try to keep that parent away from the child, while an official of the court is contacted.

If it is at all possible, in the case of separated parents the head would be well advised to try to see both of them together to agree a sensible, mutually satisfactory policy of access. For example, do the parents wish to come together or separately to school functions like open days?

Summary

The primary-school headteacher should do everything within her power to draw parents into the school in as many different ways as possible, keep them fully informed of what is happening, provide an informative school brochure, try to maintain good relationships with parents, involve them in the education of their children, and give them a voice in the making of the policies of the school.

Self-assessment

The headteacher should review her relationships with the parents of the children at her school. Does she keep them fully informed, welcome them into the school, ask their opinions and use their skills and interests?

QUESTIONNAIRE

Drawing parents in
Make a list of the methods you employ to attract parents in to your school. Check the list against the one on page 90 showing the opportunities for parental involvement. Are there any additional methods that you could use for drawing parents in to your school?

School brochure
Check the contents of your school brochure against those outlined on pages 91–92. Are you including enough information in your brochure, or could you profitably enlarge it?

Involving parents in the education of their children
What formal, organised methods do you use to involve parents in the education of their children? Do you think these are sufficient, or could you improve them?

Parents as policy-makers
How do parents have a voice in the organisation and curriculum development of your school? Could more use be made of their enthusiasm and concern?

Chapter 6
The Governors

In 1988, Kenneth Baker, the Secretary of State for Education, wrote to all school governors, stressing the importance of their role in the reorganisation of the education system:

> A central objective of our reform is to put the actual management of the school into the hands of those who know it best – the governors and the head, listening to the views of parents and staff.

The main duties of school governors are:

- To control the financial affairs of the school.
- To oversee the delivery by the head and staff of the national curriculum to the school.
- In conjunction with the LEA, to take general responsibility for the conduct of the school.
- To share with the LEA the responsibility for the maintenance and upkeep of the school buildings.
- To share with the LEA responsibility for health and safety on the school premises.
- To determine staffing levels and select, appoint and promote staff; in certain cases to dismiss staff.
- To develop a policy of discipline for the school
- To decide whether there should be a sex education policy for the school.
- To prepare an annual report for parents and hold an annual meeting for parents to discuss the report.
- In schools with more than 300 pupils, if parents so wish, to apply to the Secretary of State to opt out of LEA control and apply for grant maintained status and direct funding from Whitehall.

Appointment

The number of governors in primary schools will depend upon the number of children in the school. The table below shows how the system operates:

Pupil numbers	Parents	LEA	Head	Teacher	Co-opted (or for controlled schools foundation/co-opted)	Total
up to 99	2	2	1	1	3 (2/1)	9
100–299	3	3	1	1	4 (3/1)	12
300–599	4	4	1	2	5 (4/1)	16
600 or more	5	5	1	2	6 (4/2)	19

The parent governors are elected by parents or guardians of the children in the school. The teacher governors are elected by the teachers. The LEA appoints its representatives on the governing body. The foundation body will appoint its own governors. These governors, sitting together, will co-opt the requisite number of governors to make up the total. These co-opted members will usually represent the local business community.

It is important that the headteacher be familiar with the contents of the Instruments and Articles of Government concerning her school. These Instruments and Articles will be presented to the school by the LEA and will contain any specific local instructions for the appointment of governors.

It is the headteacher's responsibility to see that the parent governors are appointed. Like the other governors, a parent governor will be entitled to serve for four years, even if the governor's child leaves the school before the end of this period.

STAGES FOR ELECTING A PARENT GOVERNOR
Write to all parents, stating the number of vacancies, giving a brief outline of the duties of a governor. Ask for nominations for the post or posts vacant. Each would-be governor must have a child at the school and be nominated and seconded by other parents with children at the school.

Give a date by which nominations should be returned to the headteacher or any other returning officer appointed by the governors.

If the number of nominations is equal to or less than the number of vancancies on the governing body, then the nominated parents automatically become governors.

If the number of nominations exceeds the number of parent governor vacancies, an election must be held. Ask each of the nominated members to send in a brief 50-word biography and, if they so desire, an additional sentence or so about their reasons for wishing to become a governor.

Send out these biographies to all parents and accompany this list with another list of the candidates' names in alphabetical order. Send a separate list to each parent, that is, two lists to two-parent families.

Ask each parent to place a tick opposite the names or name of his or her choice for parent governor, according to the number of vacancies. If there is one vacancy the parent should tick one name only. If there are four vacancies the parent may place ticks against four different names, and so on.

Each ballot paper should be numbered. That same number should be placed against each parent's name on a list kept by the headteacher or returning officer. An envelope should accompany each ballot paper.

On a date agreed in advance, the headteacher or returning officer, in the presence of all the candidates, should open the envelopes, check the number on each ballot form against that on the list of parents, and count the votes, announcing the result.

In the event of a tie fresh elections should be held. The results of the election should be sent home to parents and announced on the school notice board.

ELECTING A TEACHER GOVERNOR
The same procedure should be followed in the case of electing a teacher governor, except, of course, that only serving teachers on the staff are allowed to vote in this case.

Helping governors to carry out their duties

The governors will not be able to carry out their duties efficiently unless they are made welcome in our schools and convinced that we value their help and advice. The most important way to do this is by encouraging the governors to come into the school.

WAYS OF BRINGING GOVERNORS INTO SCHOOL
The headteacher should set aside a specific time each week and let the

governors know that she will be pleased to see them and discuss the week's activities and problems over a cup of coffee. The hour between the end of assembly and the start of morning playtime, say 9.30 am until 10.30 am every Friday morning is usually a good time. Those governors who are not at work should be able to drop in regularly for a chat. It will help immeasurably if the chairman can get in regularly. If a running appraisal of this sort is made at the school every week it should enable most problems to be tackled before they get out of hand.

All governors should be expected to spend a morning or afternoon at least twice a term going round the classrooms, *without* the presence of the headteacher, to see what is going on and to chat informally with the teachers.

Governors should be invited to all school functions – sports days, concerts, etc. As a rule they should *not* be given special seats in the front row, but asked to mingle with the parents, getting their views in an informal way about the progress of the school.

Those governors with special skills or interests should be invited in to the school to talk to the teachers and children about them.

As a matter of routine, governors should be invited to witness or participate in any in-service training courses held for the teachers.

It is usually much appreciated if once a year the governors and staff of the school can get together for an informal social function – a wine and cheese evening, or a sherry party, or something of the sort.

GOVERNORS AND THE CURRICULUM
The headteacher can assist the governors by taking on most of the responsibility for the delivery of the national curriculum. Most governors will be pleased to delegate this function to the head, but they will rightly be aggrieved if they are not consulted in advance about any changes in the syllabuses and timetables.

The best way that a headteacher can keep the governors fully up-to-date is by preparing a series of position papers on the various core and foundation subjects and any other topics included in the school timetable. These need only contain outlines of what is going to be taught. When these position papers have been approved, they should be included in the headteacher's report on the curriculum to the governors.

This report should be a comprehensive document, updated every year and presented to each of the school's governors. In effect it contains the headteacher's philosophy and an outline of how she intends putting it into practice.

The report should cover the following topics:

- Statement of the aims of the school
- Staff details: names and duties of teachers, special responsibilities, etc
- Outline of curriculum of school
- Outline of main textbook series, teaching aids, etc. used
- Testing and recording procedures
- Marking methods
- Homework policy
- Education of children with special needs
- Education of gifted children
- Contact with secondary schools
- Staff development and teacher appraisal.

SEX EDUCATION POLICY
The governors must decide whether or not sex education should be a part of the school's curriculum. If there is to be no sex education, the governors should say so in their report to the parents. If, however, it is decided that sex education should be taught in the school, then the headteacher will be expected to make definite proposals, which could be as follows:

- The governors should consult the headteacher, parents and such interested local bodies as the health authority.
- A statement of policy should be drawn up, probably by the headteacher, for consideration by the governors.
- The sex education curriculum is probably best contained within a health education policy for the school.
- Any framework of sex education should be contained within a general policy emphasising the importance of family life and stable personal relationships.
- A working party of governors, teachers and parents should be formed to draw up the final policy, enlisting the aid of the school health service and taking into account good policy in other schools.
- The availability of existing schemes of work, videos, etc dealing with sex education in the primary school should be ascertained and examined, with a view to including them in the school's policy. The age of the children to be taught should be decided upon.
- The views and possibly the assistance of the school doctor and school nurse at a practical level should be sought.

When the final policy has been approved it should be written down and kept by the headteacher. Its existence should be made known to parents and it should be produced to anyone wishing to see it.

There is no statutory right for parents to withdraw their children from receiving sex education at school. The governors should decide whether they will allow this.

OPTING OUT

The parents or governors may decide that a school should opt out of the control of the local education authority and instead be maintained by a grant from the Department of Education and Science. This option is intended to increase parental control of a school if it is desired.

Governors of grant-maintained schools will decide upon admission policies, staffing levels and rates of payment. Like LEA and foundation schools they will still have to teach the national curriculum. At the moment a school wishing to opt out must have at least 300 children on roll, but this minimum number may well be reduced.

The headteacher of a school considering opting out may find herself in a difficult position. On the one hand she will probably have strong views about the matter, and as a governor may wish to express her opinion. On the other hand, as a salaried member of the staff she will have a pecuniary interest in the result of the decision. If the school does opt out, the governors could increase – or reduce – her salary. In the event, the headteacher would do well to consult her professional association for advice on how best to conduct herself professionally in such a situation.

One of the things for the headteacher to consider will be the change in her relationship with the local education authority if her school does opt out. There will no longer be the range of back-up and support services immediately available to her. This assistance will have to be purchased from the school's budget, perhaps from the LEA, but not necessarily. She will be free to shop around and buy in the required technical and professional advice from outside the immediate area.

Should the governors or parents consider opting out, there are certain procedures to be followed.

1. A minimum of 20% of parents must send a signed request to the governors, *or* the governors may decide for themselves to hold a secret ballot among the parents.

    In both cases, two meetings should be called by the governors, no fewer than 28 days and not more than 42 days apart.
2. A secret ballot of parents should then be held by the governors.
3. If a total of more than 50% of parents vote, and there is a majority for opting out, the governors may go on to the next stage.

4. If there is a majority for opting out but less than 50% of the total of parents voted, a second ballot must be held.
5. In the second ballot, if there is a majority for opting out, no matter how few parents voted this time, the governors must go on to the next stage.
6. Within six months of the ballot, the governors must draw up their plans for opting out of LEA control. This plan must be published and should contain:
   (a) details of existing conditions at the school and the reasons for wishing to opt out;
   (b) details of the proposed new admissions policy and the number of pupils to be admitted;
   (c) arrangements for the in-service training of newly-qualified teachers;
   (d) details of the status and character of the school under the new conditions, and details of the governing body.
7. Parents should be informed that they have the right to object to the suggestions within two months of the publication of these proposals.

PARENT-GOVERNOR MEETINGS
Each year the governors must hold a meeting to report to parents. Topics covered should include:

- Names of governors, whom they represent, when their terms of office expire
- Chairman's introduction
- Brief details of highlight of previous year's activities at school
- Report on consideration given by governors to resolutions passed at last meeting
- Details of any examination or test results
- Account of out-of-school activities – clubs, sporting events, drama, etc
- Details of financial income and expenditure
- Details of PTA income and expenditure
- Information about any arrangements for electing parent governors
- Any staff news
- Details of any links with community
- Outstanding achievements of children
- Synopses of governors' meetings over last year
- School numbers and likely intake over next year
- Social activities involving parents, staff and children

- Account of PTA activities
- Any plans for the future
- Details of time, date and place of next annual parents' meeting.

The governors may decide the time of the year at which the parents' meeting may be held. Some hold it towards the end of a school year, perhaps in June, so that an account of the previous three terms may be given. Other schools hold their parent–governor meetings in September or October, right at the beginning of the school year, and report on the previous year's activities and announce plans for the year just starting. A headteacher could try both arrangements and decide which was the most popular.

Parents should receive plenty of advance warning of the proposed meeting. Three full weeks at least should elapse between sending out the report and holding the meeting. Posters at the school gate and in local shops may help to remind people of the time and date of the meeting.

To attract parents to this meeting it is important to:

- Make the report attractive and interesting, and not too long.
- Combine the meeting with a display of the children's work.
- Show a video or filmstrip of the school's activities.
- Arrange a demonstration of music or singing from the children, as long as arrangements can be made to look after them afterwards, perhaps by hiring a video to show them under the supervision of a teacher.
- Combine the meeting with the Annual General Meeting of the Parent Teacher Association.
- Choose a suitable time, one that does not clash with a popular television programme, local disco, or whatever.
- Don't talk at the parents for too long. Make the meeting reasonably short and succinct.
- Arrange a display of items purchased by the PTA as a result of fund-raising activities.
- Announce that the head or deputy will give a demonstration lesson, using the parents as a class, to show the latest teaching techniques in a particular subject.
- Ensure that there is a break for refreshments half-way through the proceedings.
- Make sure that messages get home and that the event is well advertised locally.

- Ask for written questions in advance, but emphasise that all questions from the floor will be dealt with.
- Announce that most schools have very poor attendances for governor-parent meetings and appeal to the parents' school spirit to ensure a good turn-out!
- Announce that while the headteacher does not take part in the parent-governor meeting, she will be in attendance to answer questions if so requested.

The conduct of the parent–governor meeting will be the responsibility of the Chairman. The headteacher should give advice if asked for it, and should make sure, in a tactful manner, that the chairman is aware of his responsibilities. The room or hall should be as attractive as possible, with examples of children's work everywhere. The chairman should welcome the parents in a friendly fashion, but inform the parents that no questions about individual teachers or children will be entertained; these may be taken up with the headteacher or chairman later. It might be in order if, in a humorous manner, the question of the dangers of slander were to be raised briefly.

A quick head-count should be taken to ensure that there is a quorum of 20% of the school roll represented by the parents present, in order that formal resolutions may be passed. The chairman should then introduce each governor by name and go through the report, comprehensively but quickly. Any written questions submitted in advance should be gone through first, and then questions may be taken from the floor. Those governors with a special interest or knowledge of subjects raised should be asked to deal with them in the first instance. Finally, if everything has gone well, the chairman should propose a vote of thanks to the headteacher and staff of the school.

Governors' meetings

All governors should be expected to attend governors' meetings. These must be called at least once a term, but often more frequent meetings are needed. Seven days' notice of a meeting must be given by the chairman, the clerk or at least three members of the board.

The headteacher is *ex officio* a member of the governing body, but is not compelled to take up this position. Even if she decides not to become a governor she has the right to attend meetings. Most heads do opt to become governors, in order to have a voice and a vote in the full organisation and administration of their school.

As a rule, the headteacher should not play a prominent part at governors' meetings but should be fully briefed on the activities of the school, be prepared to speak out on its behalf and be ready to give professional advice to the governors when asked to do so. The conduct of the meeting will be in the hands of the chairman, who will be elected annually, along with the vice-chairman, by the other governors.

However, as the meetings will almost certainly be held on the school premises, the head will probably find herself involved, with the chairman, in the organisation of most of the governors' meetings.

PREPARING FOR A GOVERNORS' MEETING

Ensure that there is a clerk in position to take minutes and generally assist with the organisation of the meeting. The appointment of clerks to the governors vary from area to area. Some LEAs appoint one of the officers as clerk, sometimes the school secretary takes on the task. It is not advisable to ask one of the governors to do so; they should all be too busy taking part in the meetings.

Arrange the time, place and date of the meeting with the other governors. Sometimes meetings take place after school, but it is a good idea to schedule them just before the end of afternoon school. Governors may then be asked to arrive half an hour before the time of the meeting in order to go round the school meeting the children and staff.

Ensure that the room set aside for the meeting is warm, quiet and well-lit. Try to arrange tea or coffee before the meeting starts.

Send out reminders of the meeting, together with copies of the agenda, well in advance.

Prepare copies of the headteacher's report on school activities since the last regular governors' meeting, and send these out with the agenda.

Try to greet the governors as they arrive. If this is not possible, arrange for some of the older children to conduct them to the meeting.

Try to arrange cover for the teacher-governor's class if this is necessary.

Try to ensure that a quorum of governors is present for each meeting. This is usually one-third of the total.

The headteacher should not try to take over the meeting. Let the chairman handle it. A good chairman will be invaluable, not only for the general good that such a person can do the school, but for his or her ability to steer meetings to a successful conclusion, making sure that every subject is discussed fully and seeing that no time is wasted. The ideal chairman is always well-prepared, courteous, incisive, supportive and fair. He or she should have the gift of being able to involve everyone in the discussions and being able to arrive at a consensus. The headteacher should endeavour to build a special relationship with the chairman, based on mutual trust.

THE HEAD'S REPORT TO THE GOVERNORS
Perhaps the most important contribution that the headteacher can make to the governors' meeting is the head's report detailing the important events of the school over the past term.

This report, in essence, is an account of the headteacher's stewardship, and as such should be taken very seriously and prepared with great care. An accurate picture of the progress of the school should be provided. No attempt should be made to gloss over the problems or leave them out. The governors are an integral part of the school and should be respected and treated as such. They cannot help us sort out problems if they are not in possession of all the facts.

While there is no generally accepted format for a headteacher's report there are a number of items which should always be included:

1. *Introduction*
   Brief highlights of last three months
2. *Pupils*
   Total on roll compared with same time last year
   Totals for individual classes
   Number due to start next term or next school year
   Notable achievements by pupils
   Pupils with problems
   Any disciplinary problems
   Test or examination results
   Liaison with secondary schools
3. *Staff*
   Total staff compared with same time last year
   Changes in staff since last meeting
   Any proposed changes in staff
   Any problems with staff
   Any promotion or upgradings
   In-service training undertaken
   Any appraisal activities undertaken
   Notable achievements by staff members
4. *Main activities*
   Describe in chronological order, for example:
   16.1.90 Class 7 visited newspaper offices
   19.1.90 PTA meeting at school

23.1.90 Presentation of Road Safety Awards
5. *Visitors*
   Describe in chronological order, for example:
   18.1.90: School nurse checked children for headlice
   23.1.90: Road Safety Officer talked to school
   28.1.80 Primary Schools Inspector spent morning with probationary teacher, Mr Hargreaves
6. *Building and plant*
   State of building and plant
   Any repairs or replacements carried out
   Any repairs or replacements needed
   Any additions to building or plant
7. *Liaison with parents*
   Main activities of PTA
   Help given by individual parents with reading, etc.
   Gifts to school from parents
   Any organised efforts to bring parents into school or to visit parents at home
   Any combined staff-parent activities
   Social activities organised by parents
8. *Liaison with local industry*
   Any visits to local businesses, etc.
   Any visits to school by local businessmen
9. *Curriculum*
   Report on highlights of curriculum
   Any changes in curriculum
   Delivery of paper on any aspect of curriculum by head to governors
10. *Sports and games*
    Results of school sports events
    Any sports clubs held
11. *Cultural activities*
    Musical and dramatic performances
    Artistic displays
12. *Finance*
    Up-to-date details of school's accounts
    Details of school fund and PTA fund
13. *Future Events*
    Dates of forthcoming school activities
    Date of next governors' meeting
    Likely curriculum changes

14. *Special Report*
    Any special report called for by the governors on the curriculum or any aspect of the life of the school, for example discipline, school uniform.

Special problems

From time to time the governors will be confronted with a problem which will need special consideration. Almost invariably they will first ask the headteacher for her advice. If the head has an answer everything should be fine, but if the problem is particularly tricky she will be well-advised to take the advice of the local education authority. The headteacher should also consult the school's Instrument and Articles of Government, just in case there are any restrictions which might apply to the school alone.

The Instruments and Articles of Government have been drawn up by the LEA. There should be a copy in the school, and each governor should be given a copy as well. The Instrument is the term used for establishing the existence of the governors of a school. It provides details of the appointment of the governors, their numbers and length of office, how meetings should be conducted, and how the chairman and vice-chairman should be elected.

The Articles of Government describe the powers and responsibilities of all those concerned with the administration of the school – the LEA, the governors and the headteacher. There are also sections on the curriculum, religious instruction, discipline, and many other items of importance to the running of the school.

The headteacher and the governors must be consulted before the final draft of the Instruments and Articles of Government are drawn up for a school.

Governors' facts

SPECIAL NEEDS

The governors, through the headteacher, must ensure that children in need of special help are identified and are given the assistance they require. It is a part of the duty of a teacher to be on the look-out for children who may need special teaching, and the head should ensure that all members of the staff are aware of the needs of all children in the school.

Sometimes the needs of a child may be met from within the school, by being placed in a smaller group or withdrawn for individual help from time to time. Occasionally the assistance of special LEA remedial staff may be enlisted.

If, however, the head feels that the needs of the child are not being met by the staff or the peripatetic helpers, she must seek additional advice from the educational psychologist, doctor, or other specialist personnel.

The headteacher should have kept comprehensive records of the needs of the child and the efforts made within the school to meet these needs. Such records may be of use to the consultants brought in.

If the educational psychologist feels that much more help is needed, arrangements may be made to statement the child. This involves the official co-operation of many specialists in a formal assessment of the child's needs.

Parents should be consulted and involved at all stages of this process. The reports of the specialists involved in the formal assessment will be submitted to the LEA. The authority will then decide whether or not to issue a statement of special educational needs. The assistance needed by the child will be detailed, and the action the LEA intends taking to help the child will be stated. Such assistance could include the provision of additional aid within the school, or sending the child in a part-time or full-time capacity to a special school or unit. Assessments should be reviewed at least annually.

Summary

Unless the headteacher decides against being a governor, which is not usually a good idea, she will be responsible for involving the governors in the life of the school and accounting for her stewardship to them. She should make sure that the governors are aware of their duties and responsibilities, and organise the election of parent governors and teacher governors, or delegate this responsibility to someone else. She should write and update annually a report on the curriculum to the governors, and at each termly governors' meeting submit her report on the life of the school since the last official meeting. She should ensure that the governors draw up a policy on sex education, a code of discipline, and that they are aware of the procedure to be followed if it is decided to opt the school out of the control of the LEA. She should ensure that she or the governors presents the annual governors' report to parents, and organise the meeting at which this takes place. She should prepare for the termly governors' meeting, and be aware of the main problems likely to confront the governors.

Self-assessment

> QUESTIONNAIRE
>
> *Governing body*
> Check that you are aware of the composition of the governing body of your school in the light of the numbers given on page 110, and that you know how to organise the election of both parent governors and teacher governors for your school.
>
> *Bringing governors into the school*
> Make a list of six ways of bringing the governors into your school and involving them in the life of the school.
>
> *Headteacher's report on the curriculum and termly report to Governor*
> Check your report to the governors on the curriculum against the outline given on page 113. Can you think of any way of improving your report. Then check your last termly report to the governors against the outline given on page 119. Does yours contain most of the items given in the outline?
>
> *Opting out*
> Draw up a paper on the advantages of your school opting out of the LEA's control. Then write a paper outlining the advantages of staying within the jurisdiction of the LEA. Which seems the most convincing?
>
> *Parent-governors' meeting*
> Try to think of six ways, appropriate to your school, of attracting parents to the meeting. Check your parent-governor report against the outline given in this chapter.

Chapter 7
Finance

One great change in the functions of the headteacher in recent years has been the increased amount of financial responsibility which we have been forced to assume.

Local Management in Schools (LMS)

Headteachers have long been managing their capitation allowances in order to buy books, teaching equipment and to pay minor bills, but these sums have seldom amounted to more than 5% of the school budget.

Now that individual local education authorities have prepared their formulae for delegating the responsibility of handling most of their own funds to schools, for all but the heads of the smaller establishments, and sometimes even for these, Local Management in Schools (LMS) has become a fact of educational life.

ADVANTAGES AND DISADVANTAGES OF LMS

The advantages of governing bodies making their own financial decisions are apparent. There will be no delay in bringing in builders and other assistance. Savings made in one area may be spent in another. The extra responsibility should bring out the entrepreneurial spirit in heads and governors alike. Above all, it will give added power and responsibility to those involved in the day-to-day running of the schools.

The drawbacks to the scheme are that some heads may be apprehensive about handling large sums of money and spending more and more of their time as businessmen and accountants. Some governing bodies are worried about whether their schools are being treated fairly by the LEA when the money is distributed.

DISTRIBUTION OF MONEY

The money allocated to a school for the forthcoming year will be based on expenditure incurred in the past, the number of children on roll, and any other provisions built in by the authority. The school's governing body will then have to divide the money up among a number of different areas.

There are four main categories of expenditure:

- *Staff*
  Teachers
  Ancillary helpers
  Caretakers and cleaners
  Supply teachers
- *Building*
  Furniture and fittings
  Fuel, light and heating
  Rent and rates
- *Supplies and services*
  Capitation
  Cleaning equipment
  Office purchases
- *Running expenses*
  Printing and stationery
  Postage
  Telephone
  Travel

Against these headings will have to be set any school income from lettings, sales, etc.

OPERATING LMS

It is not the purpose of any LMS scheme to save money for a school. The intention of delegating the expenditure of the school's allowance to the governors is to benefit the children by allowing the staff and governing body to decide what the priorities of the school are in any given year.

By transferring expenditure from one heading to another, or by deciding, for example, not to paint the outside of the school, there could be savings at the end of a year which could be used for the benefit of the school as the governors see fit.

There are also economies which may be made, like using the telephone at off-peak times, switching off lights when they are not needed, and other examples of good housekeeping which should result in savings at the end of the year.

Savings should not be made at the expense of the efficiency of the school. There may be a temptation, for example, to save money under the salaries heading by employing only teachers at the lower end of the scale. Every school needs its supply of young teachers, but a head who appointed staff purely on the grounds of economy would soon end up with the sort of school that she deserved!

Fund raising

The need for extra cash is always with us, and is likely to remain so. With a modicum of goodwill from all concerned, and a certain amount of organisation, it should be possible to engage in a number of fund-raising events without disturbing the routine of the children at the school.

It is worth bearing in mind the following points before starting to raise funds:

- Develop a reputation for giving value for money.
- Raise funds for stated objectives. Parents and friends prefer to know that the money they raise will buy a new computer, or whatever.
- Contact the LEA to make sure if a licence or special permission will be needed for a particular activity.
- Advertise your functions thoroughly.
- Delegate – don't try to do too much yourself!

There are a number of fund-raising activities which can usually be relied upon to bring in a good return if they are properly prepared and advertised.

THE SUMMER FETE
Location   School field or playground, and school building.
Date and time   Summer term. Saturday afternoon may attract casual visitors, but straight after school on a Friday afternoon will ensure that older children can set up stalls, and parents can join in as they finish work.
Preparation   Advertise with posters and handbills. Place small advertisement in local newspaper. Telephone local radio station. Obtain services of volunteer helpers in advance. Get small change from bank. Organize raffle, announce that it will be drawn at fete. Hire loudspeaker.

Attracting crowd Arrange a number of 'come-ons', including pet show, athletics events for parents and children, inter-street tug-of-war competition, cookery contest for parents and children, fancy-dress event, decorated bicycle competition (remember to arrange judges for these).
Opening ceremony Not advised. Local dignitaries do not attract crowd. National celebrities charge too much.
Keeping people in Provide plenty of food and drink, children will not have been home for tea, parents may come straight from work. Organise hot dogs and ice-cream provided by local firms who will give school a percentage of profit. Run a tea and cakes stall and soft-drinks concession.
Stalls As many as can be devised: darts, skittles, bowls, etc. among the various competitions; white elephants, fruit and vegetables, tins and bottles on sale.
Extra attractions See if local small roundabouts, swings, bouncing castles may be hired, but check for safety. Sometimes youth organisations will request space at fete for their own stalls and games. Allow this, but charge a percentage of profit made!
Displays Invite local firms to set up advertising displays, and charge for the space given them.
Duration 3–4 hours.
Clearing-up Have dustbins and bin-liners on display. Give prizes for children who collect the most rubbish afterwards.
Contingencies Be prepared to move indoors if it rains!

CHRISTMAS FAIR
Location School hall and classrooms
Date and time Late in Autumn term, directly after school on a Friday afternoon.
Preparation As for Summer Fete
Attracting crowd Charge admission but have 'lucky number' on one ticket or programme, bearing a prize for the owner.
Keeping people in Have display of singing or drama from children. Provide food and drink.
Stalls As for Summer Fete
Games Any which can be performed in enclosed space.
Duration 2–3 hours

OTHER FUND-RAISING EVENTS
Sponsored Events Children are sponsored by parents, relations and

friends in such activities as spelling, singing, silence, gardening, walking, cleaning up school, etc.

Collecting waste paper Some firms pay for large quantities of waste paper, which children can bring to school. Make sure that there is space at the school for storage and that there is a guaranteed outlet for the paper.

Selling Christmas cards A number of firms will pay a school a substantial percentage of the profits if the school sells its Christmas cards and similar material to parents and friends. This needs careful organising, meticulous record-keeping, but may be taken over by PTA.

Bring-and-buy sales These events almost run themselves. Parents and friends bring in goods and give the school a share of the profits.

Car-boot sales The school could charge up to £5 a time (less if space booked in advance) to allow car to park in school playground over weekend and sell goods. A licence and permission of the LEA will usually be required.

Disco-dances These are usually popular with parents if held once a term, perhaps towards the end of term. A disco will need to be hired and refreshments provided. A licence will be needed if alcohol is sold.

Cheese and wine parties These are usually easy to organise and attract a good crowd.

Country and Scottish dancing sessions If a good 'caller' and suitable instrumentalists can be found these often provide a change from the more orthodox disco sessions.

Sports day Sell programmes, with a prize for the one bearing a lucky number. Have teas and other refreshments available.

Raffles These are lucrative, but a licence will be needed, returns must be completed, and any special requirements of the LEA complied with. There should be no cash prizes.

Bingo sessions These usually attract a loyal following, but enquiries will have to be made to see if the governors and LEA agree to gambling on the premises. If the sessions are held for the school fund and not for private gain it is probable that no duty need be paid.

School tuck shop Another lucrative proposition, selling sweets and crisps at playtime, but many teachers and parents feel strongly that sweets should not be sold at school.

School photographs One of the easiest money-raisers. A professional photographer specialising in schools' work will take the children's portraits once a year for sale to the parents, and give the school a percentage of the money gathered in.

Tee-shirts Tee-shirts bearing the school badge or logo may be ordered in bulk in certain sizes from specialist firms and then sold to the children at a small profit.

Quiz competitions   Parents and children form small teams – say two adults and two children per team – for trivial pursuits-type quiz competitions held in a large room or school hall. Admission prices are charged and prizes given to members of winning team.

The school fund

Most schools operate a school fund. Some have separate PTA funds operated by the parents. A delegated member of the staff should look after the accountancy for the school fund.

Whatever the feelings of a headteacher and her staff about having to supplement the school's income by raising additional funds it is a fact of life that most schools would find it difficult to maintain their standards without regular infusions of funds from outside.

GUIDE TO ESTABLISHING AND RUNNING A SCHOOL FUND
Discuss with the staff and PTA the exact purpose of a school fund. It may be decided the money is to be used for educational equipment like textbooks, or televisions. The exact reason for the existence of the fund must be established in advance, to avoid disputes later.

Keep most of the money in an interest-bearing bank account, transferring the money to a cheque account only for specific purposes.

Have a petty-cash box on the school premises, securely locked, for incidental expenses.

Arrange with staff just what purchases they are allowed to make, and ask for detailed accounts of their expenditure.

Keep meticulous accounts. Store all receipts.

Have the fund audited once a year, publish the results on the school notice board, present one copy to the governors, another to PTA. Announce all major income and expenditure in governors' report to parents.

There are different ways of keeping accounts for the school fund. Some are quite elaborate, and the more details which can be recorded the better it will be. At the very least the headteacher should keep a separate book in which to record all transactions. The left-hand page could be devoted to income, the right-hand page to expenditure.

Financial facts

ALLOCATION OF LEA BUDGET
A local education authority will retain a small proportion of its budget for administrative and inspectorial purposes. Of its general schools budget, 25% will be devoted to special needs, help with premises and the protection of the curriculum in small schools.

The remaining 75% will be distributed to schools. Schools will then have control of that part of their budgets covering staff salaries, upkeep of premises, rates, books and equipment and general running costs.

The amount of money allocated to each school by the LEA will be determined by formula funding. Most of the distribution of funds to a school will be based on the number of children at the school, weighted according to age. The LEA will set out the conditions and requirements within which governing bodies must operate. The authority will also monitor the performance of schools and give advice or take corrective action if necessary.

Summary

Local Management of Schools gives the headteacher and governors greater autonomy in the running of their schools, but the need for most schools to raise their own funds still remains. In both cases strict and accurate accounting methods must be adopted.

Self-appraisal

The headteacher should make sure that her accounting methods are simple and that they are scrutinised once a year by an outside auditor. Fund-raising will remain an important part of the headteacher's task. Are we doing enough in this direction?

QUESTIONNAIRE

Local Management of Schools
Do you know what your school running costs are? Try to find out from your LEA?
If you can discover the sums spent on different areas of the school's budget, work out if you would use the funds differently if you were controlling them.
Set down in order of priority the main needs of your school. Try to find out how much money will be needed to cover these areas.

Fund raising
How many of the fund-raising activities described in this chapter do you use? Can you think of any other methods of raising money for the school?

Chapter 8
Building and Plant

The LEA and the governors share the responsibility for the control of the school premises, but with the increasing financial responsibilities of the latter they will assume more of the burden of the upkeep of the building and plant. In most cases this task will be delegated to the headteacher.

The school building and the ethos of the school

If we are to present an attractive picture of the school to the local community we should start with the physical condition of the building. It stands there as a glowing example or mute reproach. Some buildings may be more attractive or modern than others, but we can all do our best to make the most of what we have.

We cannot leave everything to the school keeper or the cleaning service. No matter how efficient they may be they will need the constant encouragement and back-up of the headteacher if the premises are to remain relatively free of litter and grafitti.

This assistance should amount to rather more than the archetypal picture of the head wandering around vaguely, picking up coats and hanging them on pegs, or switching off lights in corridors. It is important, for a start, to try to make the children aware and proud of their school and want to contribute to its appearance.

They will be more inclined to do this if they see the staff taking a similar interest in their surroundings. Teachers' desks which resemble a corporation rubbish-tip, and a staff-room and headteacher's office which could pass for four-ale bars will hardly persuade the children that the head's homilies on the importance of being tidy are anything more than a typical example of the double standards of some adults.

Ways to make the building reflect the ethos of the school
Ensure that the school and its surrounds are kept clean, by liaising with the caretaker and cleaning staff and regularly checking on the physical state of the building.

Make the cleaning staff realise that you appreciate their efforts, by going round the school with them, and ensuring that their equipment is adequate.

Emphasise the importance of the work of the cleaning staff by making mention of their efforts in the brochure, newsletters and reports to governors.

Organise attractive across-the curriculum displays on a central theme, involving all teachers, changing this display regularly.

Keep a record of the school's activities over the years in a series of photograph albums on display in the school foyer.

Celebrate outstanding school achievements – successful sports teams, dramatic productions, etc, in framed photographs on the walls of the halls and corridors.

Make sure that the school is well sign-posted, with directions to the headteacher's and secretary's officers, caretaker's store, etc. prominently on view, and the names of the teachers on the doors of the classrooms.

Have a notice board at the school gate for copies of newsletters, examples of children's work, etc.

The building

There is never enough money available for the upkeep of the school building. The headteacher, in consultation with the governors, has to decide upon a system of priorities for the maintenance of the building.

Before the governors of a school take on the responsibility for its maintenance they should inspect the building thoroughly with the headteacher and draw up a list of all repairs needed. The LEA should be asked to undertake these repairs before the school takes over responsibility for its own administration.

Safety factors
Most headteachers will spend a great deal of time worrying about the safety of the children and staff. It is advisable to ask the local Safety Officer to inspect the school regularly and issue a written report containing his recommendations.

The headteacher is responsible for ensuring that the school is kept safe. Particular care should be paid to cleaning, especially of the toilets.

There should be no obstructions in the corridors. Any hazards awaiting repair should be fenced or roped off and a large warning notice put up.

There should be regular fire practices and the staff and children should know exactly where to go should there be a fire alarm. Detail of procedures to be followed should be posted all over the school.

Fire Precautions

Make sure that all safety doors are unlocked at the start of the school day.

See that there are no obstructions near any of the exits.

Make sure that the local Fire Department checks on the location and state of the fire extinguishers regularly.

See that all members of the staff know how to use these extinguishers.

Trespassers

Many people visit a school in the course of a year and most of them have a right to be on the premises. However, should a visitor prove to be a nuisance or the cause of a disturbance, then the headteacher has the right to withdraw her permission for the visitor to be in the school. Should that person refuse to go, he or she becomes a trespasser, and the police may be summoned to ask the visitor to leave.

Use of School Premises

Schools may be used for various purposes after school hours and in the holidays. The LEA and the governors decide who may use the school in this way. The LEA may have a scale of charges for non-educational organisations wishing to hire the premises.

Governors may delegate responsibility for the letting of the premises to the headteacher, but she should always check with the chairman of the governors if she has any doubts about the suitability of any group wishing to use the school building or grounds.

Licensing of School Premises

A number of public performances will generally take place on most school premises during the course of a school year. It is the headteacher's task to discover which of them will require a licence. It is advisable to consult the LEA in advance about this. The headteacher is also responsible for ensuring that the safety and sanitary arrangements are adequate.

The functions that generally do not require a licence are:

- Any performance or entertainment put on by a school, even if a charge is made, as long as the audience consists of children, parents, families, friends and invited guests. Tickets for these events should not be on sale to the general public.
- Musical performances included in a religious service or meeting.
- Garden fetes, bazaars, athletics events and displays.

Those that do generally require a licence are:

- Public dancing or musical events
- Public sporting exhibitions of the nature of boxing or wrestling displays
- Functions at which alcohol is on sale.

Plant

All equipment in the school should be checked regularly by the headteacher and staff. Contact should be made with local firms to ensure that they can come to the school speedily if any urgent repairs are needed.

Summary

The headteacher is responsible for the condition of the school building and the equipment housed in it, and for the security of the school. Crime prevention and fire prevention officers should be invited to check the premises regularly. Expensive portable equipment should be locked away securely. All doors and windows should be locked at the end of the school day. Portable equipment should be marked for security puposes. Regular fire-drills should be held and all children and members of the staff should know what to do in case of fire.

Self-assessment

It is only too easy to become accustomed to our surroundings and not to notice when they are showing signs of wear and tear. We should make a conscious effort to look at the condition of our school as it would seem to a visitor. Is the building clean and free from grafitti? Is it well-signposted? Does the display provide an attractive picture of the ethos and activities of the school? Is the safety of the staff and children properly assured? Is there an organised system for hiring out the school premises and checking that the appropriate licences have been obtained? Are harmful chemicals and dangerous equipment securely locked away so that the children cannot get near them? Are all areas of the school visited at varying times to make sure that there are no unauthorised visitors on the premises?

> QUESTIONNAIRE
>
> *Display*
> Check your arrangements for display against those given in this chapter? Could you improve upon your arrangements for making exhibitions of the children's work reflect the ethos and activities of the school?
>
> *The building*
> Are you aware of the major defects in your school building? Make a list, in order of priority, of the improvements you would like to see made.
>
> *Plant*
> What are your main deficiencies in plant and equipment? Make a list of the extra equipment you would like.
>
> *Hiring the premises*
> If you have not done so, draw up a check-list of your arrangements for hiring out the school premises to outside bodies.
> Check that you are aware of your LEA's policy for hiring and licensing school premises.

Chapter 9
The Curriculum

The 1988 Education Reform Act places an obligation upon the governors and headteachers of primary schools to ensure that their schools follow a broad and balanced curriculum which will enable their pupils to develop their spiritual, moral, cultural, mental and physical qualities and prepare them for the opportunities, responsibilities and experiences of adult life.

The Act decrees that this may best be done by teaching a number of basic subjects in the primary school.

Core subjects	*Foundation subjects*
Mathematics	History
English	Geography
Science	Technology (includes design)
	Music
	Art
	PE

Religious education must also be included as a part of the school's curriculum.

The Secretary of State has the power to lay down programmes of study for all subjects of the national curriculum. Attainment tests will be designed for children at the ages of seven and eleven.

It is to be left to the schools to decide how they will teach these subjects and how much time to devote to them. It is estimated that some 70% of the total time available will be needed to cover the core and foundation subjects. The subjects may be taught individually or as integrated studies in project and topic activities.

Designing the curriculum

Primary schools have been given by the Department of Education and

Science, a set of targets to be reached by the children in the basic subjects and the headteacher must decide how best these targets may be attained, taking into account the strengths and weaknesses of the teaching staff, the abilities of the children and the resources available.

It would be useful to have a form for subject guidelines for the school filled in by the headteacher after consulting the national attainment levels and discussions with staff:

                       Subject: ....

1st Year Infants   Term 1
                            Term 2
                            Term 3

2nd Year Infants   Term 1
                            Term 2
                            Term 3

3rd Year Infants   Term 1
                            Term 2
                            Term 3

CHECK: ATTAINMENT TESTS AT AGE SEVEN YEARS

1st Year Juniors   Term 1
                            Term 2
                            Term 3

2nd Year Juniors   Term 1
                            Term 2
                            Term 3

3rd Year Juniors   Term 1
                            Term 2
                            Term 3

4th Year Juniors   Term 1
                            Term 2
                            Term 3

CHECK: ATTAINMENT TESTS AT AGE ELEVEN YEARS

Of course, any set of attainments or targets will apply only to the average child in the class. The teacher will still have to set and monitor separate programmes for the gifted child at one end of the scale, and the one with special needs at the other extreme.

Children with special needs

It is the policy of the government as far as possible to integrate children with special needs into mainstream schools. Such children will first have to be 'statemented', ie their needs assessed by specialists employed by the LEA, before they are admitted into schools. Primary schools accepting handicapped children or those in need of special help will need to negotiate in advance with the LEA for extra resources, additional staff, and any necessary alterations or additions to the school building.

It is the responsibility of the governors, through the headteacher, to make sure that appropriate education is provided for such children once they have been admitted, that the teachers know of the needs of the pupils, and that they are aware of the importance of identifying and helping pupils with special needs.

Sex education

If the governing body decides that sex education should form a part of the curriculum of the school, the headteacher and staff will have to decide how to approach the matter. A written statement describing the policy must be issued, and all sex education must have regard to moral considerations and the value of family life.

One of the most widely used methods is to include topics on birth and growth among animals and plants in the science syllabuses for infants and lower juniors, and then to include sex education in a project on the Family in the last term for leavers in the primary school.

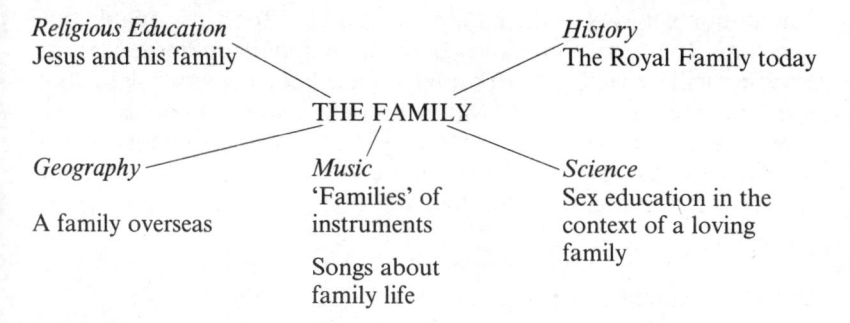

Religious Education
Jesus and his family

History
The Royal Family today

THE FAMILY

Geography
A family overseas

Music
'Families' of instruments
Songs about family life

Science
Sex education in the context of a loving family

Religious education

Each LEA must establish a local standing advisory council on religious education (SACRE). This council must draw up a suitable RE syllabus for its schools. Governors of county and controlled schools must ensure that religious education is provided for the children attending their schools and that it must be according to the syllabus provided by the SACRE, except for children withdrawn by their parents. Aided and special agreement schools have a duty to provide religious education according to the trust deed of their school, except for those children withdrawn by their parents.

Assessment and testing

Children will be tested at 7, 11, 14 and 16. Assessments in the primary school should have special regard to variations in children's early educational experiences, the demands on primary class teachers, and the need to integrate assessment into the normal teaching process.

Assessment should be criterion-referenced, showing what children have achieved against specific attainment targets, rather than in competition with one another. The tests should be formative, helping teachers to assess the needs of their children. Gradings should be moderated in order to obtain comparable standards across pupils, teachers and schools. The testing process should be related to the educational development of the children, giving continuity and progression to assessments throughout a pupil's school career.

Summary

The headteacher, at the direction of her governors, should study the LEA's curriculum policy and modify it to fit the aims and needs of her school, when necessary. The school should be able to teach the core and foundation subjects of the national curriculum for a reasonable time each day. The curriculum policy of the LEA and that of the governors, together with schemes of work used in the school and any syllabuses in the school should be made available to any parents who wish to see them, together with all statutory instruments, circulars and administrative memos relating to the curriculum. Arrangements made by the LEA for complaints about the curriculum should also be available.

Self-assessment

It is important that the spirit of the national curriculum is observed, and not just the letter of the law. Are the children in our schools being provided with an education which is challenging, balanced and capable of developing them to the full?

THE CURRICULUM

QUESTIONNAIRE

Core and Foundation Subjects
Is about 70% of the school's timetable being spent on the core and foundation subjects? Are there any changes you need to make?
Does the school curriculum receive adequate treatment in the school brochure, together with information on the school's sex education, Religious Education and how the curriculum is organised?

Religious education
Is a broadly-based, mainly Christian religious education being delivered to the children, with adequate reference to other faiths?
or
In certain circumstances, having assessed the needs and make-up of the school population, has permission been sought for the school to opt out of delivering a mainly Christian-based religious education syllabus?

Other subjects
Is time being found to include domestic science, health education and computer studies in the timetable?

Children with special needs
Does the school have a policy for the education of children with special needs?

Chapter 10
Public Relations

Publicising a school does not come easily to most heads. Nevertheless, it is one of our responsibilities. It does not matter how hard we may have worked at all the other aspects of our job if we construct a barrier of non-communication around our school. We have a duty to let the world outside know what is going on. If, through our efforts and those of our colleagues, we have managed to achieve our aim of securing that transitory and amorphous reputation of having developed a 'good' school, it is incumbent upon us to make its activities known. The prospect of going out and beating the public relations drum on behalf of our institutions can seem embarrassing and even slightly unprofessional, while the sheer mechanics of such an operation appear daunting.

The fact to be faced is that we are living in a market-place economy. Our school should be as much a part of the community as the church, the bank and the supermarket. Our neighbours have as much right to know what is going on in the classrooms as they do to be informed of the times of the church services, the financial opportunities provided by the bank, and the range of goods available in the store.

Publicising the school among the local community

We usually start with a considerable advantage here in that the community *wants* to be proud of its primary school, just as it hopes that the local football team will do well. If the school disappoints these expectations then it will be a long time, and probably another headteacher, before it recovers its place in the neighbourhood's estimation.

If, however, the district's primary school not only seems to be doing its best for the children, but also adopting a friendly and positive 'open door' policy, then this fund of goodwill should be increased. It is up to the headteacher to devise ways of involving the school in the community, and vice versa. This may be established in a number of ways.

DEVELOPING THE ETHOS OF THE SCHOOL
The behaviour and appearance of the children, the attitudes and dress of the staff and the general atmosphere of purposeful activity in the classrooms and corridors, together with a playground in which the children actually play instead of struggling for survival, will all contribute to the ethos and reputation of the school.

If the children are generally polite and well-behaved inside and outside the school environs, if they arrive in the morning looking keen and interested and leave in the afternoon in a reasonably orderly manner, most members of the community will be satisfied that all is well.

None of this will be achieved without considerable effort on the part of the staff. The expectations of the head and the teachers will have to be constantly repeated to the children. The response to these exhortations will need constant policing.

Teachers may on occasion have to be urged to take their place on playground duty a little more expeditiously then they may like. This is important. If children realise that there are certain members of the staff who may be relied upon to go on working in their classrooms or to lurk in the warmth of the staffroom for the first five minutes of each break period, then there will always be some boys and girls who will take advantage of this lack of supervision to practise a little bullying or gentle anarchy.

One member of the staff should be deputed on a rota system to take out a cup of tea or coffee to the colleague on duty, thus doing away with the necessity of teachers calling in to the staffroom on their way out to the playground. It goes without saying that all primary-school heads should take a regular part in playground duty and should set an example of promptness and willingness to chivvy pupils out of the remotest nooks and crannies. From time to time the head should take the opportunity to visit the teacher on duty, not with any intention of snooping, but to alleviate the tedium of the occasion, unless of course a visit from the head is regarded as being even more boring than standing staring into the middle distance above the heads of the playing children.

Under no circumstances should children be left unsupervised in classrooms during the break period. Those who are too ill to go out or who have work to finish should all be shepherded into the hall, or somewhere similar, and the teacher on duty should station herself where she can keep an eye on these children. When the weather is too bad for the pupils to go out at playtime the class teachers should stay with them and the head should make the rounds to allow each one in turn time to fetch a coffee.

Once the children realise that they are always being supervised, even when out of the classroom, the improvement in their standards of behaviour should be marked. There should also be basic rules about the progress of children about the school during school time – no running in the corridors, and so on.

A regular occasion when the parents and passers-by see the children in the school context is at home-time, when parents and friends gather at the school gate or outside the building. No one expects boys and girls released from the classroom at the end of the day to emerge in a decorous crocodile, but nor do they expect them to treat the space between the classrooms and the school gate as a battleground every afternoon. Consideration should be given to placing a teacher on duty at the gate for ten minutes every afternoon. Failing that, the headteacher should seek to earn a reputation for appearing among the homeward-bound boys and girls when least expected.

Technically the headteacher is not responsible for the children once they have left the school premises, but it would be a poor head and an unresponsive set of parents who did not combine to ensure that the children conducted themselves reasonably at all times. The school should never punish a child for something done off the premises, but if there is a complaint about such behaviour, the headteacher ought to contact the parents and ask them to come in and discuss the matter.

SCHOOL UNIFORM

The question of whether a primary school should adopt a uniform of some sort is a vexed one and often the cause of considerable dispute within a primary school. The arguments are well known. Advocates of uniform say that it is simple, cheap, gives the children a sense of corporate identity, and advertises the school in the neighbourhood. Opponents of the idea declare that it is old-fashioned, dull, restrictive and inhibits the expression of personality.

Perhaps the most sensible approach for a headteacher to take in this matter is to circulate the parents and ask their opinion. 'Uniform' in today's parlance usually means a combination of colours rather than a more rigid interpretation of the term. A letter along these lines might be written:

Dear Parents,
        The governors would like to know if parents are in favour of the adoption of a form of school uniform for the children. This would consist of a combination of colours to be selected by a committee of parents and teachers, for example:
        grey trousers or skirt
        red pullover
        white or grey shirt or blouse

Would you kindly complete and return this form. You need not sign it unless you wish to serve on the committee to design the form of uniform, should one be adopted.

School Uniform

If a suitable, inexpensive form of uniform, agreeable to most parents, could be devised I would be in favour of a uniform for the children ...

or

I am not in favour of a uniform for the children ...
        Please tick one of the above
Any suggestions as to the form the uniform should take
If you would be interested in serving on the committee to select the uniform, should one be adopted, please sign below.

Signed ....

CONTACTING HELPERS

The reputation of the school will be enhanced if parents and other adult helpers are welcomed to the school and informed exactly what they will be doing. This will best be accomplished by giving each helper a general letter of welcome and individual fact sheets showing them how they can be of assistance:

Dear Parent,
        Thank you very much for offering to help in the school. Your assistance will be much appreciated.
    You will be shown where all the necessary equipment is kept. The stock cupboard is next door to the headteacher's office. The sports equipment is kept in the hall.
    There are first-aid boxes in the hall, the main corridor and the staffroom.
    One of the teachers will have told you how she would like you to help with her children, and you have agreed on the day and time you will be coming in.
    The children have been told, of course, that they must treat you with

respect and courtesy. The class teacher would appreciate it if you would tell her if any of the children are naughty. We would prefer it if you left the choice of punishment to her.

Thank you again for your kind offer of help. If you are in the school at playtime we hope that you will join the headteacher and other parents for coffee in the parents' room.

Yours sincerely,

Members of the staff should also prepare and duplicate fact sheets for distribution to helpers, summarising what the teachers have already discussed with the helpers. These fact sheets could cover such areas as helping with games, drama, swimming, cooking, and so on.

ATTRACTING VISITORS INTO THE SCHOOL
Invitations into the school need not be restricted to parents alone. Favourable pubicity may be obtained by opening the school to the neighbourhood occasionally. Visitors could be attracted by:

- Organising 'fun' competitions and sporting events for adults and secondary-age children.
- Using the school as a centre for police community projects and other crowd-gathering events.
- Organising flower shows and pet shows and allowing the use of the school premises for these events.

A headteacher already reeling from the pressures of the job might regard this list with horror, but I have seen every item on it in operation at primary schools, and these were usually the schools most highly regarded in the neighbourhood, albeit with very tired headteachers!

MOUNTING DISPLAYS OF THE SCHOOL'S WORK
The activities of the school will reach a larger audience if now and again the headteacher organises a display of the children's work at various local centres. These could include the church, the community centre, the supermarket and the library. The displays could consist of a series of photographs of the school's activities – What We Have Been Doing at the School This Term – together with paintings, essays, computer print-outs and models all devised by the children. Sometimes the display could be on a theme – Autumn, Shops, etc., or a general collection of the work of the children, labelled and described in some detail.

Occasionally a local store or business organisation might be persuaded to exhibit the work of the children in its window, especially if the drawings and models exhibited are relevant to the work of the organisation concerned. A building society might like to give up some window display to a model of the buildings in the High Street, and so on.

TAKING PART IN LOCAL EVENTS
Schools will often be invited to take part in local shows and displays. This involvement might consist of helping man a float at a carnival, provide a country dancing exhibition at a fete, entering children for events at an athletics tournament, etc. Whenever time prevents this happening an opportunity for publicising the school has been lost. A headteacher should always attempt to respond in a positive way to invitations of this sort, and to make sure that the quality of the entry is up to the usual standard of the school. The children will benefit from the chance to exhibit their talents before a wider audience, and the school will receive good publicity, as long as its name is prominently displayed or announced at the function.

PARTICIPATING IN COMMUNITY PROJECTS
As a part of the social development of the children as well as reinforcing the school's reputation for being a responsible and caring institution, the headteacher should endeavour to respond to suitable appeals for help within the local community. Work of this nature must be something young children can safely carry out if carefully supervised, and a shrewd headteacher will ensure that it links up with curricular activities. Tasks under this heading might include cleaning a corner of the churchyard, or helping to clear a section of the footpath near the school.

ATTRACTING SPONSORSHIP
One of the advantages of developing a good reputation locally is that businesses in the area might be willing to sponsor the school in various ways. This may range from providing objects to be raffled at school functions, donating prizes for the children, or even subsidising the school's stock of art paper or something of the sort. In my own case the football team of a nearby public house had a whip-round and donated a complete football strip for the school team.

It will pay the headteacher to cultivate friendships with shop-owners and senior personnel in local businesses. This may be done by inviting them to visit the school, talk about their work to the children, present prizes at school ceremonies, or by asking them to provide technicians and other skilled people to assist with voluntary building and maintenance schemes at the school.

ADVERTISING SCHOOL FUNCTIONS

One simple and effective way of keeping the name of the school before the public is by advertising all major school functions by means of posters displayed in local shops and centres.

If a member of the staff is gifted in this direction she could be asked as a part of her directed time to produce the posters, perhaps with the assistance of some of the children at an after-school art club.

It will help always to use the same contrasting colours. This will identify the school in the minds of the members of the community. Even before they read a poster they will know that it comes from the school.

These hand-drawn posters can be just as effective as printed ones. They should be drawn with spirit based felt markers, using large letters, an unfussy design and a terse, simple message. They should also be taken down as soon as the relevant function is over. It looks sloppy to have large posters still advertising events long after they have finished.

USING THE LOCAL MEDIA

Local newspapers are usually keen to publish articles or news stories of interest about the schools in their area, especially if they have been carefully prepared and angled for the publication in question.

Most newspapers have a reporter specially designated as schools correspondent among his other duties. The head would be well advised to get to know this journalist and develop a bond of mutual trust.

Headteachers sometimes feel that nothing of interest to the outside world ever occurs at their schools. They underestimate the interest of the public in the minutia of local events and the desire of editors to fill their newspapers. Parents, relatives and friends are always interested enough in the activities of children at the school to buy extra copies of editions containing news about them. The headteacher merely has to sort out the right items and learn how to set them out for the press.

Possible newsworthy stories about school could include:

- School activities – concerts, plays, parties, meetings
- Parent-Teacher Association functions and meetings
- Announcements by school governors
- School trips and outings
- Unusual or well-known visitors to school
- Community service functions performed by school

- School sports events and results
- Stories about individual children or members of staff
- Items about former pupils of school

These items may be telephoned in to the newspaper office in question, or a reporter could be asked to visit the school. Most local newspapers prefer to have the stories sent to them in the form of press releases. For all stories, however, the newspaper should be given plenty of notice if the news item rates a photograph, so that the photographer can visit the school at the right time.

These points should be remembered when preparing a press release for a newspaper:

- Use A4 paper and type on one side only
- Do not use more than 250 words for a single story
- In order to get all the facts into your story remember to ask yourself the journalist's traditional questions while marshalling his material – who? what? when? where? why? how?
- Use short sentences and simple words
- Summarise the whole story in your opening paragraph, then write several more paragraphs giving details. Include several quotations from the person or people involved. In the last paragraph emphasise the role played by the school in the story, preferably through some personal anecdote
- Check every fact at least twice

Local radio stations also like to include short items of information about local schools in their news programmes, and the same sort of press release may be sent to them. Where radio is particularly interested in offers of help is in the area of speech or song. If the school choir is particularly good, or a member of the staff has written a song for the children, inform the local radio station. The manager may feel that the item is worth sending out a recording van to the school or inviting the children in to the studio to sing.

If the headteacher or a member of the staff is gifted in that direction, it might be possible to write and produce a short 15-minute or 30-minute radio play with songs which the local broadcasting service might like to record and put out over the air. A radio station will be particularly interested in short plays for children which have a local connection or which are capable of being broadcast at particular times of the year – Christmas, Easter, Harvest, and so on.

To interest a local television station an item of school-based news will have to be especially interesting or relevant to persuade a producer to send a cameraman and recording engineer to the school. However, most local television news services are on the lookout for 'fillers', short human-interest stories to fill the odd gap in the schedules. Interviewing children on a relevant theme is highly favoured here. My own school was once visited by a crew from Yorkshire Television and the children in the first-year infant class were asked whether they really believed in Father Christmas. If a headteacher writes to the local television station with a list of suggested topics in this manner, the broadcasters may not use them but they will make a note of the name of the school and the fact that the headteacher is prepared to co-operate in such ventures.

Sometimes the attention of the media may be less welcome. If something has happened which does not reflect well on the school, or which may be controversial, then reporters may telephone the school asking for details.

If it seems likely that this may happen then the headteacher should act quickly. The chairman of the governors and the local education authority should be consulted, to make sure who should handle any such enquiries. A detailed press statement should be written and agreed. If the head is to field any approaches from the media and she is approached by a reporter before the statement is ready, she should take the telephone number of the journalist and promise to ring him back and read him the passage. This promise should be kept. A headteacher should not ignore the attentions of the press; its members will only contact other informants and probably come up with a garbled version of the truth.

When the head has read the press statement to a reporter, she should politely refuse to enlarge upon it or to answer any supplementary questions. If the journalist wants to know more he should be referred to the chairman of the governors or the local education office.

Publicising the school over a wider area

Very few of us are going to have the virtues of our schools extolled in *The Times Educational Supplement* or made the subject of a BBC feature on *Panorama*. In fact hardly any primary school is ever going to achieve more than a local reputation, and this should be enough for most of us.

A headteacher who sets out expressly to conduct a public relations exercise on a national scale will usually be wasting her time. Local organs may be only too delighted to publish news about the school but national newspaper and television companies are much more hard-nosed in their approach. Every day these institutions receive hundreds of glossy press hand-outs from professional public relations companies. The activities of one distant primary school are hardly likely to interest the tabloids, unless something dreadful or salacious has happened there.

Those schools who do achieve regional or national fame do not as a rule do so because the headteacher has set out to achieve this. In almost every case such schools have received publicity because of outstanding efforts on the part of the staff and children, or because they have come up with an original or amusing idea and translated it into action. In all the cases I have known of schools of this type, they have been amazed to find their efforts publicised on a large scale.

A study of those primary schools whose activities have aroused favourable interest on a national or regional scale reveals that their activities include educational successes in major poetry and literature festivals; sporting and cultural distinction in such areas as cricket, chess and drama, and unusual and amusing ideas, of the order of dinner supervisors being issued with red and yellow cards to be brandished at recalcitrant children, in the manner of football referees. These are some examples of primary schools attaining publicity:

- St Cutherbert and St Mathias Primary School, Earls Court, for developing an effective method of multicultural, multi-faith education in a school where the average length of stay for a pupil is 18 months.
- St Stephen's Primary School, Bradford, for its liaison activities with a neighbouring special school.
- Bawdsey Primary School, Suffolk, for its skilful deployment of staff and resources in a small school.
- Grove Junior School, Handsworth, Birmingham, for involving parents in the life of the school.
- Foxhill Primary School, Sheffield, for devising a home–school reading programme.
- The Abbey Hey Junior School, for getting its recording of *The Sparrow* into the Hit Parade.
- Melcombe Primary School, Hammersmith, for the headteacher's idea of carrying a walkie-talkie with her as she walked around the school, thus keeping in touch with her secretary.
- Llanybydder Primary School, Dyfed, Wales, for some of its children devising a magnetic walking stick to attract keys and other metal objects dropped by its owner.
- Ochiltree Primary School, Ayrshire, for winning a national compe-

tition sponsored by *The Times* for producing the best school newspaper, despite having only 98 children in the school.

Public relations with the local education authority

The relationship of headteachers with officials of their local education authorities have undergone a number of significant changes over the years.

Once there was the rather remote and impersonal 'office', with little contact with schools. This was followed in the 1970s by an injection of funds and enthusiasm, leading to the creation of many schemes, some of them excellent, others less so, involving the local education authorities and their schools. Then in the 1980s came the cutbacks and a lessening of contact between officials and headteachers.

Now we are entering a new era, with all sorts of opportunities and challenges. LEAs will become more streamlined as additional duties devolve upon schools. There will still be opportunities for joint projects, but they will have to be carefully chosen. How these chances are taken up will depend upon the amount of co-operation between each LEA and its headteachers.

Unless she is in charge of one of the opted-out schools, a headteacher will still depend a great deal on the LEA for advice and assistance. It is more important than ever to build up and maintain a good, friendly relationship with local officials and support workers. In this new age bridges will have to be built, and headteachers will play a prominent part in their construction.

This may be achieved by such obvious measures as responding with courtesy and reasonable efficiency to requests for information, filling in forms on time, conscientiously complying with all local regulations, and attending such meetings as may be called by the authority. Heads should keep local officials in touch with what is going on and invite them to take their part in activities and relevant decision-making policies of the school. For their part it is to be hoped that the LEAs resist all temptations to adopt a siege mentality, and show an awareness of the problems of headteachers.

The most common point of contact with the LEA will be through the local inspectors or advisers. Informal contact with these colleagues can at the very least lead to a headteacher being kept up-to-date with what is going on in the d... ion, and at best, thanks to the 'pollination' principle, receiving information about good practice in neighbouring schools. The best inspectors work side-by-side with teachers in the classroom, helping them to develop professionally.

PUBLIC RELATIONS

The inspectors are also the eyes and ears of the LEA. If our schools are doing well the inspectors will relay this fact, accomplishing a valuable public relations function for us. If we are doing less well, they should have the skill and experience to help us put matters right.

At a more formal level, a school might become the subject of an official inspection or visit, by a single inspector or a team. Sometimes a head might request such a procedure, perhaps to help particular members of the staff, or test curricula strengths and weaknesses, or just to get an idea of how the school seems to be doing to informed outsiders.

As a rule, however, the initiative for an inspection will come from the LEA. There should be prior consultation with the head, at which certain ground rules should be established. The inspectors should let the head know what the reason for their visit is, and who will be making the inspection. At this meeting the headteacher should also request that the inspectors make an oral report on their findings, so that these may be discussed before their report is written.

A forthcoming inspection will always worry teachers and headteacher alike. We all want to look our best. The head should do her best to reassure her staff, and point out that the school should benefit from the expertise of those conducting the inspection.

Most inspectors are far too shrewd and experienced to be taken in by any form of 'window-dressing'. They will, however, expect a head and staff to be concerned enough about their school to want to present it in the best possible light.

It may, perhaps, be a fitting conclusion to this book, intended as it is to help heads develop the potential of their schools, to present a brief outline of some of the qualities that other professionals may look for and hope to find in them.

SOME AREAS OF EXCELLENCE POSSIBLY SOUGHT BY INSPECTORS MIGHT INCLUDE:
- Quality and state of maintenance of grounds and buildings
  playground   gardens   playing fields   parking facilities
  classrooms   library   TV room   staffroom   secretary's office
  headteacher's office   toilets   medical room or facilities
  cloakrooms   stairways (if any)   caretaker's room
  resource areas   quiet areas for study
  science and design and technology areas   hall   corridors
  waiting areas for visitors   play areas   school meal areas
- Decoration and display

state of paintwork and fabric generally
quality of display material   signposting
- Quality and deployment of staff
ratio of teachers to children   distribution among classes
range of experience, expertise and recent training
appraisal arrangements   distribution of incentive allowances
curricular and pastoral responsibilities
out-of-school activities
arrangements for and conduct of staff meetings
ease of contact with headteacher
opportunities to gain management experience
degree to which non-teaching staff members are integrated as a team with teachers
special responsibilities and opportunities for deputy
- Attitude of children
general ethos   positive attitude to discipline   pastoral care
ability and opportunities to work independently   happiness
sense of purpose   self reliance   courtesy   confidence
trust in adults
degree to which individual needs are met   degree of bullying
acknowledgement of all talents, not just academic and sporting
ability and willingness to participate in out-of-school activities
willingness to accept responsibility and help others
pride in school
- Headteacher
organisational abilities   leadership qualities
relationships with staff, parents, governors, children
teaching load   part played in community
degree of accessibility to all
ability to absorb pressure
- Organisation
admission policy   range of classes   group work   class sizes
contact with parents   reports   newsletters   homework
children with special needs
chance for teachers to work with different age groups
range and balance of work outside and inside classroom
arrangements for transfer to and liaison with secondary schools
organisation of secretary's work
- Equipment and resources
books   musical instruments   games and PE equipment
audio-visual aids   television and radio sets
arrangements for ordering and replenishing stock   reprographic-

equipment   video recorders   computers   stage lighting
amplification equipment   art and craft equipment
science and design and technology equipment
- Curriculum
guidelines for all core and foundation subjects
guidelines for other subjects
policy for assemblies and religious education
'Match' of guidelines and what is actually happening
part played by entire staff in drawing up guidelines
constant updating and assessment of curriculum
flexibility of timetables
amount of expertise brought in from outside school
- Parental involvement
presence of PTA or other parents' group
arrangements for parents helping in school
social activities involving parents
arrangements for parents seeing children's work
arrangements for parents to see teachers at short notice
arrangements for keeping parents in touch with what is going on at school
degree of satisfaction with school expressed by parents
- Involvement of governors
frequency of formal and informal meetings of governors
regularity of meetings between chairman and headteacher
frequency with which governors visit school, meet staff and children
contribution made by governors to curriculum development
degree of general involvement of governors in life of school
degree of satisfaction with school expressed by governors
quality of governor/parent report
- Assessment and record-keeping
thoroughness of assessment and record-keeping
degree to which all child's qualities, not just academic, are recorded
way in which prizes, if any, are presented, reflecting all areas of school life, not just academic
perceptiveness of comments in reports
detailed knowledge shown of all circumstances of child's home-life
way in which test results and samples of children's work are recorded and stored
quality of marking procedures

- The school and the community
  sense of community shown within school
  amount of commitment to value of individual shown
  links with local community–businesses, church, etc.
  links with individual members of local community.

Summary

The most effective form of good public relations is to develop a school which is popular with parents, the community and the children because it provides an interesting, relevant and comprehensive form of education in the widest sense of the word.

In 1989, Professor Peter Mortimore of Lancaster University conducted a study of 2,000 children in 50 London schools and concluded that the sort of school to which a child was sent made a considerable impact on the progress the child made, and provided a checklist for parents wishing to test the efficacy of the education being provided at a school. The main qualities of a good school were:

- head and teachers with a consistent philosophy
- staff fiercely ambitious for *all* pupils, whatever their ability
- head on good terms with staff and respected by pupils
- an 'open door' policy for parents
- examination results appropriate for the school's social intake
- homework set and marked
- informal and formal methods of teaching used as appropriate
- firmly maintained discipline.

As headteachers we should see to it that the community gets the opportunity to find out what is happening in our schools. In order to do this we should invite people into our schools, make sure that the layout of the school and the behaviour of the children attract favourable notice, take a full part in neighbourhood activities, let the local media know what we are doing, and do our best to make our schools as effective as we possibly can.

Self-assessment

Most headteachers do not like publicising their schools, but it is a necessary part of our job. We should ask ourselves if the school is running well enough to stand examination, and then do our best to let the community know what is going on. Are we making the most of our school building? Is the display inside attractive? Is the behaviour of the children good inside and outside the school? Do we make the most of our opportunities to take part in local events? Do we play a full part in the service of the community? Do we use the different media in our area to let people know about our successes?

QUESTIONNAIRE

The school and the community
Draw up a plan for welcoming members of the community into the school. How many different functions can you list which will draw in visitors?
Try to involve the school in at least one aspect of local community services.
See if you can persuade at least three local firms to sponsor a school activity.

The school and the media
Find out the names of the reporters and producers responsible for dealing with schools on your local newspaper and radio station.
 Draw up a news story about something which has happened in your school, in accordance with the suggestions given in this chapter. Submit it to a local newspaper.

Further Reading

The Headteacher's Job

Peter Drucker, *The Practice of Management*, Heinemann, 1954.
Valerie Hall, Hugh Mackay and Colin Morgan, *Headteachers at Work*, OUP, 1986.
Michael Marland (ed.), *School Management Skills*, Heinemann, 1989.
Derek Waters, *Management and Headship in the Primary School*, Ward Lock Educational, 1979.

School Organisation

HMI, *Ten Good Schools*, HMSO, 1977.
HMI, *Primary Education in England*, HMSO, 1978.
HMI, *Primary Schools: Some Aspects of Good Practice*, HMSO, 1987.
ILEA, *The Junior School Project*, ILEA, 1986.
Martin Leonard, *The 1988 Education Act*, Blackwell, 1988.
Tom Marjoram, *Assessing Schools*, Kogan Page, 1989.
Ken Reid, Robert Bullock and Stephen Howarth, *An Introduction to Primary School Organization*, Hodder and Stoughton, 1988.

The Children

Stephen Adams, *A Guide to Creative Tutoring*, Kogan Page, 1989.
Mary Drummond, *Four Year Olds in the Primary School*, Falmer Press, 1989.
Maurice Galton and John Willcocks (eds.), *Moving from the Primary Classroom*, Routledge, 1983.
John McGuiness, *A Whole School Approach to Pastoral Care*, Kogan Page, 1989.

FURTHER READING

Florence Maltby, *Gifted Children and Teachers in Primary Schools, 5–12*, Falmer Press, 1984.
Tom Marjoram, *Teaching Able Children*, Kogan Page, 1989.
Patricia Marshall, *Transition and Continuity in the Educational Process*, Kogan Page, 1988.
A. Pollard, *The Social World of the Primary School*, Holt, Rinehart and Winston, 1985.
K. Reid, *Truancy and School Absenteeism*, Hodder and Stoughton, 1985.

The Staff

N. Bennett, *Teaching Styles and Pupil Progress*, Open Books, 1976.
D. Boydell, *The Primary Teacher in Action*, Open Books, 1979.
M. Galton, Brian Simon and Paul Croll, *Inside the Primary Classroom*, Routledge, 1980.
Geoffrey Hall, *Teachers, Parents and Governors: their Duties and Rights in Schools*, Kogan Page, 1989.
Anthony Smith, *Starting to Teach*, Kogan Page, 1988.
Suffolk Education Department, *Those Having Torches: Teacher Appraisal: a Study*, Ipswich, 1987.

The Parents

B. Goacher and Margaret Reid, *School Reports to Parents*, NFER/Nelson, 1983.
Alistair Macbeth, *Involving Parents*, Heinemann, 1989.

The Governors

Tyrell Burgess and Anne Sofer, *The School Governors' Handbook and Training Guide*, Kogan Page, 1986.
Ann Holt, Norman Rea and Neville Price, *The Governor Training Pack: Getting Started*, BBC Enterprises, 1989.
Chris Lowe, *The School Governor's Legal Guide*, Croner, 1988.

Finance

Peter Downes (ed), *Local Financial Management in Schools*, Blackwell, 1988.
Gerry Gorman, *Fund-raising for Schools*, Kogan Page, 1988.
B. Knight, *Managing School Finance*, Heinemann, 1983.

The Curriculum

David Alexandra and Anne Pennell, *Implementing Science and Technology in the Primary School*, Falmer Press, 1989.

Len Almond, *The Place of Physical Education in Schools*, Kogan Page, 1989.

R J Campbell, *Developing the Primary School Curriculum*, Holt, Rinehart and Winston, 1985.

Colin Connor, *Assessment and Testing in Primary Schools*, Falmer Press, 1989.

W S Fowler, *Towards the National Curriculum*, Kogan Page, 1988.

Geoffrey Hall, *Records of Achievement: Issues and Practice*, Kogan Page, 1989.

HMI, *The Curriculum from 5 to 16*, DES, 1985.

Douglas Newton, *Making Science Education Relevant*, Kogan Page, 1988.

Kogan Page Books for Teachers series
Series Editor: Tom Marjoram

Assessing Schools Tom Marjoram
Fund-raising for Schools Gerry Gorman
A Guide to Creative Tutoring Stephen Adams
Making Science Education Relevant Douglas P Newton
The Modern Primary School Headteacher Graeme Kent
The Place of Physical Education in Schools Len Almond
Practical Teacher Appraisal Diane Montgomery and Norma Hadfield
Records of Achievement: Issues and Practice Geoffrey Hall
Starting to Teach Anthony D Smith
Teachers, Parents and Governors W S Fowler
Teaching Able Children Tom Marjoram
Towards the National Curriculum W S Fowler
Transition and Continuity in the Educational Process Patricia Marshall
A Whole School Approach to Pastoral Care John McGuiness

Index

access to children 107
adapting to circumstances 24, 28
admission of children 30
adults in school 89, 90, 95
advertising for staff 62
advertising school functions 126, 148
aims of school 16–17
ancillary helpers 82
applications for posts 63
appraisal of headteachers 75
Arnold, Dr 18
aspirations of headteacher 16
assaults on staff 85–6
assemblies 37–50
attendance 106–7
attitudes of staff to head 34, 71, 72

Baker days 78–9
basic subjects 25
bereavements, dealing with 105
brochure 91–2
building, school 132–3
bullying 45–6

capitation 124
care and supervision 47, 57
caretaker 83
cascade method 76
chairman of governors 118
class organisation 52–3
communication with parents 91–6
communication with staff 70–2
community projects 147
composition of school day 23–4
computers 92–3
conditions of employment 83
consulting parents 102
consistency 35
contact with children 32–3
contact with colleagues 70–2

core and foundation subjects 25
custody of children 107–8

Data Protection Act 30
delegation 68
deployment of teachers 67
deputy headteacher 81–2
designing the curriculum 137
detention 57
diaries, school 93–4
directed time 21
discipline 46
disciplinary procedures 47
discrimination 65
dismissal of teachers 84
displays 146
disruptive children 43–4
duties of headteacher 14

educational psychologist 44
Education Welfare Service 44
educational visits 50–1
equipment 135
ethos of school 143
exclusion of pupils 44
extending experience of staff 67–8

fairs 127
family grouping 53
fetes 126
financial headings 125
fire precautions 49
first aid 49
Fox Hill Reading Project 99
fund raising 126

gifted children 54–5
governors and the curriculum 112
governors, appointment of 110–11
governors, duties of 109

INDEX

governors' report to parents 115–17
GRIST 78
grouping 52–3

headteacher's report to governors 119
home reading schemes 98
homework 100

ILEA Junior School Project 16
image of headteacher 11
incentive allowances 68
INSET 78
interviewing parents 95
interviewing staff 64–7

job descriptions 62

lesson preparation 74
letters 94
licences 128
local radio 149
log books, children's 97
log of headteacher's time 27

media 148–50
merit system 42

newsletters 92
newspapers 148
non-directed time 22

objectives of headteacher 11
open evening 103
opting out 114–15
organisation of time 21–4

paperwork 152
parents as policy makers 101
parent-governors meetings 96, 115
parent governors 110–11
parental involvement 96
parental responsibilities 44, 106–7
parents' room 91
parent–teacher associations 104–5
playground duty 143
police 56–7
posters 148
premises, safety of 49, 133
premises, use of 134
preparations for secondary school 55
pre-school children 89
pressure 11
press releases 149
probationers 79–81

punishments 43

qualities of a good school 16, 153
qualities of headteacher 18
questionnaires 102

Race Relations Act 65
Reading Partnerships in Hackney 98
record keeping 51
records of achievement 52
recruitment of teachers 61–2
references 63
registers of children 56
relationships with colleagues 72
religious education 38
reports on children 93
respect 41
responsibilities of headteacher 15
responsibilities of teachers 49, 68
rewards 42

safety of premises 133
school day 23
school fund 129
secretary 83
self-directed time 22
Sex Discrimination Act 65
sex education 113, 139
special needs 139
sponsorship 147
staff meetings 71
staff newsletter 72
staff, non-teaching 82–3
standards of school 142
stress 11
supervision after school 57
style of headteacher 12

targets for teachers 75
teacher-governors 111
team teaching 67
telephone, use of 94
Ten Good Schools 12
timetable, school 26
Tizard, Professor J. 98
training of teachers 76–8
treatment of teachers 69
trespassers 134
truancy 46
types of headship 12–13

vertical grouping 53

'whole' school 19